About the A

JOHN NIELSEN is an environment correspondent for National Public Radio and a fourth-generation Californian who grew up in the condor's rangelands. As a reporter, he has come to specialize in stories about endangered species and changes to the natural landscape. He lives in Washington, D.C.

CONDOR

CONDOR

═══

To the Brink and Back—

The Life and Times of

One Giant Bird

═══

JOHN NIELSEN

HARPER ● PERENNIAL

NEW YORK ● LONDON ● TORONTO ● SYDNEY

HARPER ●○ PERENNIAL

A hardcover edition of this book was published in 2006 by HarperCollins Publishers.

HarperCollins books may be purchased for educational, business, or sales promotional use. For information please write: Special Markets Department, HarperCollins Publishers, 10 East 53rd Street, New York, NY 10022.

FIRST HARPER PERENNIAL EDITION PUBLISHED 2007.

Designed by Joseph Rutt

The Library of Congress has catalogued the hardcover as follows:
Nielsen, John.
Condor : to the brink and back—the life and times of one giant bird / by John Nielsen.—
1st ed.
    p.   cm.
ISBN-10: 0-06-008862-1
ISBN-13: 978-0-06-008862-0
    1. California condor.   I. Title
QL696.C53N54 2005
598.9'2—dc22

                                                                2004054336

ISBN: 978-0-06-008863-7 (pbk.)
ISBN-10: 0-06-008863-X (pbk.)

07 08 09 10 11 ❖/RRD 10 9 8 7 6 5 4 3 2 1

For Danny, Jonah, and Eli

CONTENTS

ACKNOWLEDGMENTS

This book was made possible by people who offered me more help than I had a right to ask for, by people who helped me through some long dark nights, and by people who know the California condors better than I ever have or will.

Thanks to Megan Newman, my original editor at HarperCollins. Thanks as well to my endlessly patient agent, James Levine, of the Levine Greenberg Literary Agency. Thanks to the entire staff of the science unit at NPR News; thanks especially to Anne Gudenkauf, NPR's science editor. In the darkest of those long dark nights, I got infusions of support from the wise and powerful Nancy Huddleston Packer of Stanford University. Barton Kraff, Kathleen Nadeau, and Nancy Ratey also helped to keep me sane. Naturally the most important late-night aid and comfort was provided by my brother, Peter, my sister, Kirsten, and my parents, Tom and Marilyn Nielsen.

Researchers and archivists and human packrats had a hand in the making of this book. They include Kee Malesky at NPR News; Susan Snyder at the Bancroft Library at the University of California-Berkeley; John Heyning of the Los Angeles County Museum of Natural History; Jan Hamber of the Santa Barbara Museum of

Natural History; Linnea S. Hall of the Western Foundation for Vertebrate Zoology; Steven Herman of Evergreen State College; Dave Phillips of the Earth Island Institute; Les Reid of Pine Mountain, California; Greg MacMillan of Shandon, California; Jon Borneman of Ventura, California; and Antony Pietro of Santa Barbara, California. I also got important help from the Library of Congress, the California Academy of Sciences, the Ventura County Museum of History, and the Santa Clarita Valley History Society.

Every living leader of the field teams who worked to save the condor took my calls and shared their secrets. So did the leaders and condor teams at the Los Angeles Zoo and the San Diego Wild Animal Park. Special thanks in that regard to Lloyd Kiff, Noel Snyder, Mike Wallace, Bill Toone, Shawn Farry, Sophie Osborne, Robert Mesta, Sandy Wilbur, Mike Scott, Mark Hall, John Ogden, and Fred Sibley.

Last and most I'd like to ask you all to rise and cheer the unsung field grunts still at large in the California condor's domain. They're the ones who'll save or lose this species in the end and they need to know that we're behind them.

INTRODUCTION

The California condor is the Elvis Presley of endangered species. It is huge, it is iconic, it is worshipped and despised, it is beautiful and hideously ugly. It does a wicked mating dance and eats colossal meals. And, it's not really dead.

The California condor is a New World vulture with telescopic eyes, a razor-sharp beak, and a wingspan of nearly ten feet. Helicopter pilots say they've seen it soaring well above ten thousand feet. I have seen it glide for miles without ever bothering to flap. Condors never show the wobbly V shape you see on smaller, lesser vultures. The giant black wings form a horizontal line that's often mistaken for an airplane. The thick black feathers at the ends of these wings are nearly two feet long.

Condors used to travel in the company of birds with wings that made their own wings look weak. During the Pleistocene epoch, these birds formed a scavenging armada that searched out the carcasses of mastodons. Meals the size of cars were quickly reduced to piles of gristle and bone. Then, ten thousand years later, the other giants vanished, leaving the condor alone. Ever since, it's been the largest living flying thing over North America.

The condor is the soul of the wilderness. The condor is smarter than you think. The condor is a rat with ten-foot wings and the enemy of progress. It's a bird whose sad demise and incomplete recovery is a preview of the future facing lions, tigers, bears, and other charismatic species.

Take what you want from the paragraphs above, but be forewarned. Once this bird gets into your head, it does exactly as it pleases. You can try to shoo it away, do it harm, and forget it's even there, but the condor will still be there in the background somewhere, biding its own sweet time. Then one day it will rise, spread its giant wings, lean into the wind, and own you.

I remember standing near the edge of the cliff in the Hopper Mountain Wildlife Refuge in Ventura County, California, one day in the 1980s when I heard a distant rumbling sound rise up the side of the cliff: it was so strange and strong that it didn't seem to need a shape. Whatever it was that was making that sound clearly didn't care who or what might hear it coming.

As the rumble turned into a roar, I took a half-step toward the cliff, thinking that sound was vaguely musical. That was when the mammoth shape exploded up past the lip of the cliff into my field of vision: an adult condor with its wings spread wide, shooting straight up into the air.

That's the condor that owns me now. It's not the one I grew up dreading, a bird that entered my consciousness in the early 1960s when my father rented a mansion at the edge of a virtual ghost town less than fifty miles north of downtown Los Angeles. The town once known as Piru City was built in the 1860s by a man named David Cook, who'd made a fortune selling Bibles and religious tracts in his home state of Illinois. Mr. Cook had brought his weak lungs west for some clean air and bought a portion of a cattle ranch. On this land he'd planted rows of citrus trees and fruits mentioned in the Bible. Then he invited nonsmoking, nondrinking white Christian families to come and live in his "Second Garden of Eden," picking fruits and vegetables in a wholesome, healthy setting. To lure the pickers in, he built a Main Street and a beautiful Methodist Church. For himself he built a three-story Queen Anne mansion in the hills behind the town.[1]

Then, unexpectedly, Mr. Cook sold his land to an oil company and took his fully recovered lungs home to Illinois. Some people say Mr. Cook left town because the trains kept bringing drinkers and smokers to Piru City. When he left, the town began decaying. It was still decaying in the 1960s when my father took a job in the area and started looking for a place to put his wife and kids. Mr. Cook's seven-room Queen Anne mansion was for rent for $160 a month. I thought it looked spooky, but Mom and Dad liked it: one week later we were in.

I was a child of the suburbs then—Dad had come to build one— and now we were in a house that fit that world about as well as Dorothy's shack fit the Wonderful Land of Oz. Mr. Cook's mansion was a creaking relic of the 1880s, with stained-glass windows warped by the sun, termite-infested twenty-foot ceilings, and a red stone tower with a turret. An abandoned outhouse in the backyard was full of wasps and spiders, and the ruined wood reservoir a little farther back was full of lizards and scorpions. My mom fell in love with Piru instantly. Little brother Peter, baby sister Kirsten, and I were a little stunned at first, but in about a week it was home.

Dad drove east to a truck stop called Castaic every day on a road so narrow and full of curves that it was known as Blood Alley. His office, a former ranch house, was full of architects and builders who'd been hired to plant a city on a swath of empty grazing lands and feedlots. It was to be called Valencia, after the metropolis in Spain. This Valencia was to be a "planned community" that featured split-level ranch-style homes and centrally located shopping, all with easy access to a brand-new interstate.

My brother and I went to Piru Elementary School, where rattle-snakes sometimes crawled down from the hills and onto the football field. When school was out, we ran through the orange orchards. When we got home, we climbed up into the turret at the top of the tower. To the south were orange orchards and the sandy bed of the lazy-looking Santa Clara River. Past the river, farther south, ran a line of low, steep mountains. Beyond them another freeway cut down through Simi Valley, pulling malls and suburbs in behind it.

Behind the turret to the north was a scarier set of mountains, taller and more forbidding than those to the south. Dad told us once that far beneath these mountains, big slabs of rock called tectonic plates were fighting with one another and pushing the mountains up from below. Sometimes the fighting got so bad that the mountains split and crumbled as they got pushed higher up into the air. Dad said there were giant cracks called fault lines in these mountains. When they moved, the earth shook. Sometimes people died.

The mountains behind Piru are part of a set of ranges that used to separate Southern California from the rest of the state. For at least a century, these mountains were feared by travelers passing through them: they were where the bandits and the fugitives hid out, as well as the biggest, most ferocious grizzly bears.

The last of the California condors made their homes in this forbidden zone, raising their young in the caves of the cliffs. For all I knew, the birds were there to guard the hidden castle of some kind of evil queen. I even went out looking for that castle once or twice, climbing through the yucca and the thistle on the steep, dry hills directly behind the mansion. I took naps in the shallow caves in the wall of the dried-out ravine. I found a tarantula one time and took it to third-grade show-and-tell, and on another occasion, I wet my pants in terror when a rattlesnake under a bush decided to make his presence known. I even started a brush fire once that could have burned the mansion down.

When I was exploring, I kept an eye out for the monster birds. A Mexican kid named Mike Ortega gets both the credit and the blame for that. He was the one who swore that gringo-eating vultures always hid inside the storm clouds, waiting for kids like me to stop moving. If I did, the condors would come shooting down like lightning bolts, ripping me to smithereens, or maybe carrying me back to the castle of that queen. Mike's dad, Angel Ortega, later told me that his son was fibbing: what condors really did was kill and eat eight-year-old kids who played hooky from school and sneaked around in the orange orchards, smoking cigarettes. I was almost sure this wasn't true, but I was not about to take chances, since at the time I was a hooky-playing eight-year-old who smoked cigarettes in the orchards. For a long time I looked back for giant black wings up at the edges of the clouds. But I never saw them, so I stopped.

═══════

Twenty years later, I got a call from an editor at *Sports Illustrated*, asking whether I would like to write a story about a pair of doomed vultures living in the mountains north of Los Angeles. They were the last two free-flying condors in the world, he said. Roughly twenty more were living in off-exhibit breeding pens at two Southern California zoos. Federal biologists had been sent in to catch the last two free-flying condors and take them to the breeding compounds. Some environmental groups supported the plan, while others were absolutely livid. One of the angriest critics was the legendary David Brower, who saw the condor struggle as a land war. Brower thought the condor needed wilderness like a fish needs water: when the birds were trapped they would no longer be condors, he argued, and the wilderness would be badly wounded.

The people who ran the condor program said Brower didn't know what he was talking about, adding that the wilderness was no longer a safe place for condors to be. They didn't know why, but they knew the birds were dying out there, and so they had to take the final step.

I agreed to do the piece and drove back to Piru, which had yet to emerge from its coma. After meeting with trappers and their critics, I wrote a quick-and-dirty guide to the California condor wars, in which I attempted to explain how the species had reached its sorry state. By the time *Sports Illustrated* published the story in early 1987, there was only one wild condor left, and that one didn't last very long. On Easter Sunday that year, the condor known as AC-9 (or Igor, to his friends, perhaps because he walked in a stiff-legged, clumsy way when he was a chick) was captured and transported to a zoo.

This was one of the sadder days in American environmental history, but it may also have been the day the condor was saved. Condors did well in captivity, or at least that's how it looked. In 1992, a pair of birds that had been hatched and reared in zoos was released in the mountainous back country of Ventura County, near the place where Igor had been captured. Many more releases followed, as the zoo-bred condors were restored not only to the Transverse Ranges north of Los Angeles County but also to the coastal mountains near the town of Big Sur. In 1996, more zoo-breds were set free near the Grand Canyon. In 2003, they were released in the mountains of Baja California. From a possible low of fewer than twenty in the late 1970s, the zoo-based condor population had grown at least tenfold, to the point where there are now more than two hundred of the birds alive, and for that I am grateful. The monster of my youth had been returned to me.

But present and accounted for is not the same as saved. It could take a hundred years or more to truly save this species, and no one thinks salvation is assured; its grazing lands are full of carcasses shot full of lead. For fifty years the condor recovery program has been a kind of scientific bar fight. On top of all the injuries sustained in the field are the shattered friendships and the grudges, and the careers that were ruined in the service of the bird. Some of these fights were inevitable, given how close the condor has come to extinction: every decision is a gamble in such a situation, and one or two mistakes could be enough to bury the species.

Many millions have been spent on condors since the end of World War II—twenty million in federal and state dollars, which makes the California Condor Recovery Program easily the most expensive endangered species conservation effort ever mounted in the United States. Every move the captive condors make in the zoos is videotaped and studied at great length, and all one hundred condors in the wild have tracking devices on their wings. The birds are trapped and tested for a wide range of toxins on a regular basis. New satellites may soon track their movements in real time.

It's true that the condor has long been a species with no ecological value. Some critics say that means we need to let it go. Other experts argue that the money spent on condors could be put to better use.

And, yes, there is that other more basic complaint: Why should anybody want to save a bird as ugly as a California condor? This is a vulture that cools its legs by peeing all over them, a vulture that jams its head into the soft parts of dead things. It's a bird that decorates the walls of its caves in layers of feces and vomit, a bird whose bald, red, and badly scarred head makes it look like the survivor of a terrible fire.

Fortunately, these images vanish the instant the bird takes flight. You may think there's no chance you could ever give a damn about this bird, but take my word for it: once you see the condor soaring, it owns you.

I don't think the condor will be saved until people take the fight to save it personally. I do, and here's why. While visiting a condor refuge near Piru, I helped some field biologists trap a dozen condors for blood tests. When the birds were back up in the air again, I saw a battered wing tag lying on the ground. It had just been taken off a condor known to some as AC-8 and to others as the Matriarch. AC-8 was one of the last free-flying condors captured in the 1980s, when she and Igor were a breeding pair. At the Los Angeles Zoo, she'd bred for years, producing many offspring. In the late 1990s, she was the first of the formerly wild condors to be reintroduced to the wild.

When I held up her tag to one of the biologists, he said, "Oh, go ahead," so I put the tag in my pocket. When I got home, I stuck it up on the wall in front of my computer, and sometimes when I couldn't write, I looked at it for a while, trying to imagine it hanging from the wing of the Matriarch as she soared the base of a storm cloud at ten thousand feet.

A few months later, the Matriarch was blown away by a pig hunter who said he didn't know the bird with the giant wings was a California condor.

I am not the kind of guy who dedicates books to a vulture, but if I were, this book would be for her.

THE WORST OF TIMES

===

Before I go on with this short history, let me make a general observation—the test of a first-rate intelligence is the ability to hold two
opposed ideas in the mind at the same time, and still maintain the
ability to function. One should, for example, be able to see things
that are hopeless and yet be determined to make them otherwise.
—F. Scott Fitzgerald, *The Crack-Up*
(Charles Scribner's Sons, 1931)

The pit traps used to catch the last wild condors looked like
shallow graves. When Pete Bloom slid down into one and
closed the small trap door, he entered a clammy earthen trench that
was six feet long, two and a half feet wide, and approximately four
feet deep. It was hard to move around down there without smashing
your head on the support beams; not moving meant dealing with
cramps that paralyzed your back, neck, and legs. Bloom said he
often passed the time lying on his back next to the walkie-talkie,
waiting for word that the last of the wild condors had arrived.

He liked it down there. He had to. In 1985, they were his office.
"Typically I went into them before sunrise and came out an hour
before sunset," he said. "Then often, I'd go back down into them
the next day."

This is what the fight to save the California condor had come
down to in 1987—buried biologists waiting for the chance to leap
up out of the ground and grab the last free-flying condor, ending
an era that had lasted for at least ten million years.

Nobody in the condor program liked that image, and some absolutely loathed it. But the scientists who knew the condors best knew they'd run out of options. Something in the condor's habitat was poisoning the birds, and rumors that somebody was trying to kill them were all over the place. The scientists were warning that a reproductive emergency was at hand. Every last condor had to be caught and brought to zoos for captive breeding.

Bloom believed the arguments held moments of weakness. "I felt like the state executioner," he told me once. "But I knew we were doing the right thing, so that's what I focused on."

In 1987, Bloom didn't look like the kind of guy you'd want to leave your kids with: he was skinny to the point of scrawniness with wild brown hair and a beard, and strange-looking scars made by talons and beaks on his leathery face and hands. He had been raised about a hundred miles south in Orange County, California, where his father maintained helicopters at the local Marine base. When he was a kid, he started trapping red-tailed hawks near his house for the fun of it, and then he started fitting them with tags that helped researchers track their movements. Over the years he'd learned to trap all kinds of other raptors, using everything from cannon nets to wire mesh baited with mice. When Bloom joined the condor recovery team in the 1980s, he was known as one of the most accomplished and reliable trappers.

Packing for the pit traps was a ritual for Bloom. Into his black filthy briefcase always went one walkie-talkie; one set of binoculars; one small battery-operated ceiling fan; one bag of lunch with an extra-large water bottle; one piece of airtight Tupperware with a small roll of toilet paper inside; one lucky hunting knife; one dirty rug; and one 100-percent-cotton sleeping bag. Synthetic bags were out because they were too noisy. Coffee, deodorant, strong-smelling foods, and bug sprays were also forbidden, even though the condors weren't thought to have a strong sense of smell. "They were avoiding us and we didn't know why," said Bloom. "I wasn't taking any chances."

The field crews dug at night when the condors were asleep. Usually it took a crew of six to build a trap from start to finish: three or four field biologists, one veterinarian, and one or two designated "master baiters," so named because it was their job to bolt the carcasses of stillborn calves to the ground in front of the trap. This job usually involved driving out to a local dairy and then wading through knee-deep pools of manure and urine to get the carcass, which was then hosed down, cleaned up, and moved to a freezer close to the trap. "Road-kill deer went in the freezer, too, if they were big enough," said Bloom. "The only thing we never used were the carcasses of animals shot and left behind by hunters."

When the trench was finished, it was reinforced with four-by-fours and covered by an inch-thick sheet of plywood. The trapdoor Bloom climbed in and out of was at the front of the structure; in the middle was a head-size hole covered by an upside-down wicker basket that was porous enough to see out of. When Bloom went in, the basket and the plywood were covered with dirt and bits of vegetation—in the end, it looked like a bump in the pasture.

Scavenging birds of every shape and size were quickly drawn to the carcasses—ravens, turkey vultures, black vultures, and golden eagles. When Bloom heard the birds hit the ground, he'd check the wicker viewing basket for black widow spiders, often squashing one or two beneath one of his boots. Then he'd push his head up through the hole in the plywood and peek at the mayhem taking place five feet in front of him. Sometimes Bloom saw a half dozen golden eagles fight for choice chunks of meat while another half dozen stood back waiting for an opening. Once he saw an eagle dive at least three hundred feet into the back of another large bird, knocking it senseless and clear of the spot the eagle wanted on the carcass.

"Ravens sometimes parted the grass in front of the basket with their beaks," said Bloom. "They would see my eyes looking at them, and back away like nothing had happened. I'm certain they knew I was there, they just couldn't believe it."

He could have reached out and grabbed any number of golden eagles by the legs: he'd done it dozens of times while working other jobs. But condors were another matter.

"Very cautious birds," he said. "Sometimes they'd fly over the carcass once and never come back, and other times they'd circle down and land on a dead branch near the top of a tree. They'd watch the other birds eat for hours and then turn around and leave. That's what usually happened."

Bloom said it was easy to recognize the sound of an approaching condor: the whoosh that became a roar kept getting louder and louder until it ended with the thump of great big feet and the clatter of enormous wings. When the condors walked directly over the trap, Bloom could hear them breathing, their wheezing lungs sounding something like a winded child's. Looking through the basket at the carcass, he would see the smaller birds start flapping and scattering about, jumping out of the condor's path like peasants diving off the road at the approach of the king's carriage.

Condors were usually trapped by nets fired out of small cannons, but Bloom tried not to use them when he didn't have to. There was always the chance that one of the cannons would fall over and start a fire, or shoot too low and blow a hole in the bird. The guns might fail to fire all at once; plus, they required explosives. Finally, it was very hard to hide a cannon.

But the pits, if built properly, were undetectable, he said. From the air they blended in perfectly with the soil and the vegetation, and when the birds were on the ground, they did not notice the difference in terrain. "Eagles and condors on the ground look up and around for danger all the time," he said. "But they hardly ever look down for predators, and we used that to our advantage."

The trapping had gone slowly and fitfully, but by the end of 1986, there were only two condors left in the wild. Bloom caught one of them, a bird known officially as AC-5, on February 27. He remembered looking up and seeing the silhouette of the last remaining wild California condor set against the clear blue sky. That bird was Igor.

━━━━━

David Brower ached to see the condors, but he did not want to track them down. They were creatures with a driving need to stay away from people, he had always said. People who handled them were biological thugs, in his opinion, "macho scientists" who sat around comparing their scars. To Brower, this was conservation science at its worst, if it was conservation at all.

"A condor is [only] five percent feathers, flesh, blood and bone," he once wrote in an open letter. "The rest is *place*. Condors are a soaring manifestation of the place that built them and coded their genes. That place requires space to meet in, to teach fledglings to roost unmolested, to bathe and drink in, to find other condors in (and not biologists), and to fly over wild and free."[1]

In 1987, this was the rhetoric that powered what was left of the "hands-off" school of condor conservation, which at one time had been endorsed by the state's most powerful and accomplished natural scientists.

But the number of condors left in the wild had fallen since the end of World War II, and drastic scientific interventions in their lives had become routine. Plans to trap the condors and breed them in captivity had now been endorsed by all the relevant government agencies and a panel of nationally known ornithologists and natural scientists.

But Brower would not be persuaded to endorse their method, and that gave people pause. He was one of the most important environmentalists in American history, famous for his hatred of compromise solutions, and this, in his view, was way past compromise. This was virtual extinction.

I called him in late 1986, while the trapping was under way: he spent the next fifteen minutes firing broadsides at the federal recovery program, using words and phrases I had heard him use before. The California Condor Recovery Program was a perfect example of how *not* to save a disappearing species: it was a high-risk experiment on a creature that deserved much better.

"To save it, the condor was destroyed," said Brower on the phone. "The zoos have everything they wanted."

Brower's confrontational style was one of the reasons he no longer worked for the Sierra Club, and it would soon be one of the reasons for his split with Friends of the Earth. But Brower's seething eloquence was his genius, too: early on it helped him define the modern American environmental movement.

Brower wrote the book on environmental land wars in the early 1950s, just after he became director of the Sierra Club. The Federal Bureau of Reclamation, long known in the West as an agency that could not be defeated, had just announced that it would raise a series of new dams in an astonishingly beautiful corner of the Colorado River Basin. Western cities and so-called recreationists loved the idea, but Brower hated it. Even before it was revealed that one of the dams would flood part of Utah's Dinosaur National Monument, he was preparing to fight it.

When the plan to flood a small part of the park was made public, Brower launched the land war that earned his place in history. To stop the dams, he organized ranchers and lawyers and hikers, who rarely spoke to each other, and hired Wallace Stegner, Utah's answer to Shakespeare, to edit a picture book/anthology called *This Is Dinosaur*. He then commissioned a documentary that demonized the bureau by arguing that the dams could not be built safely.

Then he drew the whole country into the fight by buying a full-page ad in the *New York Times*. The ad declared that there was really only "one, simple, incredible issue here—this time it's the Grand Canyon they want to flood. The GRAND CANYON." When the bureau tried to rally support from fishermen and water-skiers, Brower placed another full-page ad in the *Washington Post*: "Should we also flood the Sistine Chapel so tourists can get nearer the ceiling?"

The Sierra Club lost its tax-exempt status after these ads ran. But the bureau lost the war. Three of the proposed dams were not built, and it's all but certain that they never will be. According to the writer John McPhee in his *Encounters with the Archdruid*, these tactics set the gold standard for future environmental battles.

He went after the basic mathematics underlying the Bureau's proposals and uncovered embarrassing errors. All this was accompanied by flanking movements of intense publicity—paid advertisements, a film, a book—envisioning a National Monument of great scenic, scientific and cultural value being covered with water. The Bureau protested that the conservationists were exaggerating—honing and bending the truth—but the Bureau protested without effect. Conservationists say the Dinosaur Victory was the birth of the modern conservation movement—the turning point at which conservation became something more than contour plowing.[2]

Those were the strategies Brower carried into the condor wars, which he viewed as a fight to save not only the condor but a creature whose plight was symbolic of the threat to endangered species all over the world. "We can respect the dignity of a creature that has done our species no wrong," he once wrote. "Except perhaps to prefer us at a distance."

Brower often told reporters that his interest in condors dated to the early 1930s, when he was a daring and extremely accomplished rock climber. Then, in the 1940s, he met a man named Carl Koford at the Museum of Vertebrate Zoology at the University of California-Berkeley. Koford was finishing the study that defined the condor as a creature of the wild. Brower came away convinced that saving them meant saving wild places.

When Koford died of cancer in 1979, Brower became the alpha male of the hands-off movement. With the help of Dave Phillips, a colleague and close friend, Brower questioned every move the federal condor team made. Friends of the Earth, the group he then ran, put out a book called *The Condor Question: Captive or Forever Free?* and a documentary that showed a condor dying in the arms of a biologist.

However, when I called Brower in 1986, the battle had been lost. Lawsuits had failed to stop the trapping plans; calls to influential friends and federal officials had been ignored. The governing board of the Sierra Club had voted to support the "hands-on" approach, as had a panel of nationally prominent natural scientists. Phillips said he and Brower faced those facts near the end of 1986, while sitting in the darkened San Francisco office of Friends of the Earth. "We'd done everything we could do to stop the zoos and their friends, but it hadn't been enough. We knew the wild bird was gone," Phillips recalled. "If we got a bird back from the zoos, we'd get a tamed-down version of the condor—a bird that had been trained to fly back and forth inside a grid."

By then, David Brower had gone looking for the birds, overcoming previous misgivings. In 1984, at the age of seventy-two, he joined a group of bird-watchers near the summit of Mount Pinos, in Ventura County, where condors sometimes showed themselves. His hair had long ago turned white and he couldn't climb anymore, so he waited with the crowd near a parking lot, holding his binoculars.

"Then came my lucky day," he wrote years later. "A pair of birds moved toward us, probably well attuned to the human throng on the Mount Pinos summit and adding to their own life list of people they observed."[3]

Brower saw the large white triangular patches underneath the wings, and the big red fleshy heads on the periscopic necks.

"What those birds did for the wild sky lifted my heart and gave the vault above me a new dimension I had never paid attention to before," he wrote. "I could wonder what it would be like to think the way a condor does, to sweep so much of creation at one glance, to know the wind."

He was beside himself. "When you have waited seventy years to witness such a performance," he wrote, "when you have been willing to forgo it if the witnessing might somehow affect the condor's survival, and when the display is finally there before you, you can be excused for feeling excited, euphoric, and especially privileged."

When Brower and I spoke briefly on the phone after that, I asked him whether he'd go looking for a zoo-bred condor that had been restored to the wild. Brower said he didn't think he would live to see it happen.

Turns out he was wrong, but not by much.

———

The man in charge of the condor-breeding center at the Los Angeles Zoo had been sleeping on a roof near the breeding compounds with a rifle at his side. Mike Wallace might have heard an owl or two above the endless drone of the traffic, but on most nights that was it, unless you counted the environmental activists climbing up the trees of Griffith Park, which surrounds the zoo.

"Earth First!, Animal Liberation Front, people like that," he said. Wallace says the activists also tried to howl like wolves from time to time.

As the last few wild condors were brought to the zoos in the late 1980s, hard-line environmentalists kept hinting at plans to "liberate" the captive birds. Native Americans claimed their culture would officially die when the last wild condor was captured. At the Los Angeles Zoo, a man in a condor costume roller-skated back and forth all day in front of tourists, handing out leaflets that made it look like the zoo was an Alcatraz for condors.

Wallace thought a lot of these people were fools, but he tried to keep that thought to himself. Sometimes, he'd go out to try to reason with the activists, but those conversations went nowhere.

"Very weird times," he said years later. "The PR staff was freaked out by the bad publicity, and by the thought that something terrible would happen when AC-9 was captured. They were like, 'Oh my God, what if it dies in here?' I told them to relax. Good things were going to happen."

Wallace told the staff to expect an uproar when Igor was captured, but assured them in a few days it would all die down and the activists would go away. The birds would breed and the chicks would grow into healthy wild condors. Then, barring a meteor strike, some of the condors born at the zoo would be released into the wild, and finally the Los Angeles Zoo would begin to escape its reputation as a minor-league institution. People would describe it as the place that saved the California condor, and not as a cheap imitation of the larger and richer zoo in San Diego. Wallace believed with all his heart that this was almost certainly the way the future would play out, and when he said as much to the PR staff, it helped to calm them down.

But Wallace also knew there were outcomes to be feared. One of the fears was that the chicks produced by these birds would be deformed by a genetic disease. Or maybe the condors *would* refuse to breed in captivity; maybe their chicks would be tame. Returning condors to the wild would then be impossible. They were going to need a lot of luck at the zoos. Wallace kept that thought to himself.

"Nobody had ever done these things before," he said. "We'd been forced to make a lot of key decisions without much information. You can work your ass off to prepare for these things, but you're never really ready."

Wallace said he used the rifle he slept with on the roof of the zoo to shoot at the vermin. "Rats and foxes used to try to get to the breeding pens. I shot a few of them, no big deal. I grew up on a farm in Maine. I would never have thought of using it on the activists," he said. "But the activists didn't know that, and that was fine with me."

━━━━━

The mood was very different at the breeding center run by the San Diego Wild Animal Park, partly because it was set in the foothills of some dried-out mountains in a yet-to-be-developed part of San Diego County. Activists never climbed in the trees making unconvincing wolf sounds there. No one in a condor suit roller-skated in the parking lot.

But the park had gotten a host of threatening phone calls, and it was prepared for trouble: tall wire fences topped with concertina wire surrounded the facilities, and the guard who opened the gates checked strangers' IDs closely.

Inside the compound the rows of tall wire flight pens were covered on all sides with plywood. Fourteen California condors were living in those pens at the start of 1987, leading what captive-breeding expert Bill Toone said were quiet, uneventful lives. Like Mike Wallace, Toone had been studying condors for most of his career, preparing for the day when the future of the species would be in the hands of the zoos. He was absolutely certain the wild birds would breed.

"No doubt whatsoever," he said. "We'd spent years working Andean condors, and they were breeding prolifically. California condors were different birds, but not that different."

But Toone was also a nervous wreck in 1987. Tensions on the field teams were alarmingly high, and he was sick of being portrayed as a "condor cager," ignorant of the need to save habitats as well as birds.

Toone said he had always shared David Brower's urge to save more of the condor's rangelands, even though he didn't think there was a shortage of protected condor habitat. What he didn't share—what he detested—was the idea that the zoos had been plotting to capture all the condors for decades now. He blamed Brower and his allies for the condor decline, noting that they had snuffed a plan to breed a very small number of California condors in captivity in the early 1950s. Those would have hatched a lot of eggs, Toone argued—many more than they would have been able to produce in the wild. Those captive chicks would have grown up into condors that could have been released into the wild, said Toone. By the late 1980s, they'd have been thriving out there.

The magnitude of what it meant to take a condor out of the wild hit Toone one day in the mid-1980s when he climbed into a condor sanctuary to get to an egg a condor had just laid on the floor of a craggy-looking cave. Toone and biologist Noel Snyder were planning to take the egg to a zoo where it would be hatched and raised.

When the two men reached the cave, they noticed the mother of the egg was inside watching over it. Toone and Snyder hid and waited until the mother condor flew away. But, as soon as they entered the cave, Toone heard a vaguely musical roar approaching the cave entrance. He turned to see the mother condor swooping back and forth past the opening, hoping to make the egg thieves go away.

A wave of doubt washed over Toone at that moment. "It occurred to me that there was a living embryo in this egg—if it hatched in captivity, it would spend most of its life in a pen that was no more than forty feet wide, eighty feet long, and twenty-two feet tall, and this was a bird that soared at altitudes measured by the mile. In that instant I promised myself I'd stay with the program until a bird raised in captivity was released to the wild. Then I would walk away."

Near the end Pete Bloom had a recurring nightmare: Igor lands on the carcass and Bloom catches him easily, but then he trips and falls on top of the bird, killing it instantly. Next thing he knows he is standing in front of a microphone at a press conference, looking out at glaring lights and hundreds of reporters. The press conference is endless, but there's only one question: How did it feel to kill the last free-flying condor?

Bloom had been waiting in the clammy darkness of the pit trap for several months now, but Igor had not taken the bait. The last free-flying condor saw the carcass, and he often circled down a bit to take a closer look. But he must not have liked what he was seeing down there—after a few minutes, most of the time, he would turn and fly away.

No one in the condor program was surprised to hear that Igor was the last bird left. He'd been closely watched and tracked since the day he hatched in the wild, and he'd always been remarkably independent, if a little klutzy. "When he was young, he was a very

friendly, curious bird," said Bloom. "It was easy to approach him."
Bloom says Igor's friendly disposition held when the condor started
looking for a mate. It kept holding when Igor's first attempts to
breed went nowhere, partly because his mating dance seemed to go
on forever, and partly because he kept attempting to mount his
partners from the front.

Then in the eighties, the trapping started and Igor was never the
same. Bloom says Igor was captured twice by people hidden in pit
traps; once for blood tests, and once to have radio transmitters
bolted to his wings. After those events, when approaching a carcass,
Igor was exceptionally cautious, sometimes watching other vultures
eat for days from the top of nearby roost trees.

Bloom said he'd looked up and seen this condor looking down as
the other condors were captured. No one knows what the condor
learned by watching other birds, but Bloom had a hunch his bird
had learned to see the pit traps from the air. Maybe the mound of
earth that covered Bloom's viewing basket was the clue that gave the
traps away. Maybe it was the way the cannon nets were always buried.

Bloom also gave some thought to another explanation: maybe
someone on the trapping team was tipping Igor off. "All you'd have
to do was take a tiny mirror out and flash the sun into the condor's
eyes. You might also make a sudden movement when nobody on
the ground was looking. I was pretty paranoid for a while there."

Bloom and his crew built a new kind of trap in April 1987; the
pit and the viewing basket weren't placed so close to the carcass,
and a row of small cannons rigged to fire a big net over the carcass
was added and disguised as bushes. A long, buried black cord connected to the first cannon ran over to the pit trap and into the bottom of a plastic tube with a red button on top. Bloom held the
trigger as he waited.

Igor did a lot of wandering in April 1987, sometimes flying north along the crooked spine of California's coastal mountains, or winding west through the Transverse Ranges. ID tags with great big 9s on them hung from the front of his wings, and a radio transmitter was bolted onto the left wing. Teams of Igor trackers in pickup trucks had been chasing the beeps sent from the transmitter, but for weeks the last free-flying condor had toyed with them. Sometimes he hovered over the pickups, looking down at them. Then he'd turn and disappear over the top of a roadless mountain for several days, panicking his would-be captors.

Bloom's hopes rose on April 18, when Igor landed on a dead branch at the top of a tree near a trap. The bird stared down at the carcass for a while, but then he flew away. Bloom sent the crew home so some of them could spend the next day celebrating Easter. When Bloom was home he got a call from a radio tracker named Jan Hamber, who told him that Igor had returned to the tree and fallen asleep. Bloom immediately called the trapping crew back into the field.

Igor woke up late the next morning, as condors are wont to do. He lounged around and spread his wings to catch the sun, never leaving his perch. Several hours later, as if on a whim, he stepped off his branch and floated down toward the carcass. He landed just beyond the range of a big mesh trap net attached to four small cannons. He paused to look around for what Bloom remembers as an eternity.

And then the condor clomped past the basket and planted one of his fat pink feet on top of the carcass. After one more look around, Igor whipped his beak down, burying his head inside the carcass. For an instant, he was blind and deaf. Bloom pushed the button that fired the net.

Four bushes seemed to explode. A thin black line sailed out of the ground between the bushes, spreading out into a net as it arced over Igor's head. As the bird began to run, there was another crash, and a bearded man flew up out of the ground.

Igor, running and flapping his wings, was one step from the edge of the net when Pete Bloom caught him. With his right arm, he closed the wings; his left hand closed the beak. When someone slipped a hood over Igor's head, the last wild condor stopped fighting.

Not long afterward, Igor was locked inside a medium-size dog kennel and driven to the Oxnard airport, in Ventura County. There he was loaded into a plane that did not go to Los Angeles, where activists were ready to rattle the fences at the Los Angeles Zoo, but to an airport in a rural part of San Diego County, the San Diego Wild Animal Park. Wallace and Toone had secretly agreed to make the switch to avoid the scene up in Los Angeles. By the time the demonstrators figured it out, it was way too late, said Wallace. No demonstrations were reported.

The stories that ran in the papers the next morning read like eulogies, which seemed appropriate. No matter what happened to the birds in the future, April 19, 1987, would now be marked as a low point in American environmental history. And here's what I have always considered the saddest thing about it: of the tens of millions of people now living in the condor's former rangeland, few know what condors are. Fewer still know what the condor used to be, or why that's so incredibly important.

WING IN A GRAVE

═══

The story of our condor will be reckoned from the wondertime of North American fauna, the Pleistocene, the final epoch of the 70-million-year-long Cenozoic era. Seldom has animal life known more fascinating diversity or greater numbers than during this epoch. Men living today who have an interest in such things can but weep for not having seen it. We can know it only from the mountains of bones.

—Roger Caras, *Source of the Thunder:*
The Biography of a California Condor

Drenched in sweat and unsure of his sanity, Dr. Steve Emslie wedged his feet into a notch in the cliff and waited out the wind. Several hundred feet below him, the Colorado River twisted south through the bottom of a dark red canyon. Emslie saw the boatman who had brought him down the river standing at the water's edge. Damn, did he look *small*.

"I was wondering where my partners had gone," Emslie said, referring to the professional rock climbers he had hired to take care of him. "I was not a practiced climber, and I was not sure what to do."

Emslie was trying to get to a cave on the outskirts of the Grand Canyon National Park. In that cave, he hoped to find evidence that the Grand Canyon area had once been a condor nirvana, where the giant birds had found plenty to eat, lots of company, and many caves in which to lay their eggs. Condors like caves that can't be reached by wingless predators, and there seem to be an infinite number of caves like that in the Grand Canyon area. Some are thousands of feet above the dark green river, many are hundreds of thousands of years old, and the range of shapes they take is dazzling. There are caves with multiple entrances, caves hidden behind boulders, and caves that lead into huge caverns. Many had never felt the weight of a human foot, and a few had spooky reputations: one of the caves Dr. Emslie wanted to enter led back into a cavern that produced an eerie moaning noise.

Caves in the walls of the Grand Canyon have been dry for many thousands of years. Animals that die inside them decompose *very* slowly, to the point where bits of skin and hair sometimes hang from the skeletons of prehistoric animals. Clues to what the landscape and the weather were like when these animals lived can be found in crusty heaps in the corners: ancient pack-rat middens made of sticks, leaves, bones, and feathers, along with anything else the rats could carry up into the cave.

Emslie hoped to find the bones of long-dead condors in these caves, along with the bones of the things they ate. He wanted to know more about the lives these condors led—what they ate and where they nested. He also hoped to figure out when they had left the canyon, and if he was lucky, why. But first he'd have to get off the ledge he was on.

He was connected to the other climbers by a safety rope, but the rope was long and he couldn't see to the other end of it. Lacking a plan, he froze, felt the wind try one more time to pry him free of his perch. *Giant wings would come in handy here*, he thought.

When Emslie started up this cliff in 1986, it was known that condors had once ranged not only up and down almost all the Pacific Coast but across parts of the Rocky Mountains and along the Gulf Coast. Condor bones had been unearthed in Florida and in upstate New York. Emslie thought the first true condors had evolved in

North America about eleven million years ago, branching into the modern genus *Gymnogyps* eight million years later. He believed these North American condors found their way to South America, where they split off and formed not only a new species but a separate genus. These became known as Andeans, with slightly different markings, slightly larger wings, and a tendency to kill and eat small animals.

Then, at the end of the Pleistocene, the bird's range imploded. No one knows why or how, he says, but one thing is certain: change arrived with what geologists think of as blinding speed.

Emslie wanted to look in the caves for clues to what happened. When he began his study, the bones of remarkably well-preserved condors had already been found in some of the canyon's more accessible caves, along with Indian artifacts and the bones of other kinds of animals. Many experts thought that meant the condor hung on in these caves until the late nineteenth century, or even until the start of the twentieth century. That would put the condors in the region when the first groups of humans settled in the canyon four thousand years ago. Those birds would have seen the Anasazi culture thrive in the Vermilion Cliffs for roughly two thousand years, and then watched whatever it was that caused the Anasazi to vanish without a trace.

This was a politically important story line in the late 1980s. Rumors of federal plans to start releasing condors in the area had just begun to circulate, and many of the people who lived near the canyon weren't sure they liked the idea. Some of the most vocal critics didn't think the feds had the right to return condors to a place they'd left many thousands of years before. Lawsuits that might press that point would not get very far if proof of recent nesting could be found. Bits of condor eggshell dating back one hundred years might be enough to do the trick; condor bones that stopped appearing ten thousand years ago might be enough to stop the project entirely. What the feds needed most was proof that condors had laid eggs and raised chicks in the canyon recently, which seemed to mean sometime in the last fifty to one hundred years. Finding the bones of adult condors would not be enough, since an adult could have been transient and died while passing through. The bones of a chick might be plenty of evidence, on the other hand. Tiny fragments of a condor eggshell might also be enough.

Emslie didn't enter the condor caves to settle a political argument, but he definitely had the skills he'd need to do it. In the 1980s, he filled an esoteric niche in the ornithological world, being one of the world's experts on the bones of long-dead condors. This was a man who knew how to sort fossilized droppings by species and to identify the tiniest condor bone on sight. If the caves held any proof at all of condor nesting, he would find it—assuming he didn't kill himself on his way to the caves.

Emslie's professional hero was the late Loye Miller, who was one of the first to find the fossilized bones of condors, in the early 1900s. Miller was a dapper, genial man, known to students and colleagues as "Padre," an eloquent writer who thought of old rocks and fossilized bones as great works of art. In an essay entitled "Ornithology in the Looking Glass," he wrote of "the enlargement of the spirit" that comes "with the growing concept of ornithology as extending backward into an almost imponderable past—fauna preceding fauna, shifting with shifting scene."

That's what Emslie was chasing after when he got stuck on that ledge: bones that pointed back into this "almost imponderable past." Retracing his steps, he started edging backward, hoping the synthetic rope around his waist would lead him back to the guides.

"That was when the chunk of rock broke off above my head," he said. "It fell between my feet and smashed into the ledge, right on top of the rope. I thought, *Great, if that rope's broken, I'm a dead man.*"

Turned out the rope wasn't broken. Not long afterward, Emslie pulled himself into a cave that contained not only the bones of long-dead adult condors but the bones of a condor fledgling that had died in its nest. Scattered around the cave were bits of condor eggshell fragments and the bones of giant sloths.

"It was as if I'd entered a museum," Emslie said.

<hr>

The Pleistocene epoch was the condor's prime. The lands beneath the soaring birds were littered with the carcasses of mammals bigger than any seen before or since. Giant flying scavengers such as the condor would have led one another to these carcasses and then methodically turned the carcasses into piles of bones.[1]

It's hard to fathom how big some of these carcasses were. Consider the largest species of North American mammoth. Alive, it sometimes weighed ten tons and stood fourteen feet high, with fourteen-foot-long tusks curving around like giant fishhooks. The mammoth's smaller cousin, the mastodon, weighed only six tons by comparison, but the distance from the ground to its shoulders sometimes measured ten feet. Giant ground sloths ambling up from South America were as big as modern elephants.

There were many more of these cartoon beasts, ranging from tiny prehistoric horses to ridiculously oversize bison. Huge camels with puny humps roamed near llamas with enormous heads; caravans of deerlike creatures known as pronghorn traveled under moving fields of giant antlers.

The carnivores that killed these creatures were impressive in their own right, starting with the saber-toothed cats that used their seven-inch fangs like matching pairs of daggers, and moving right along to the carnivorous bears with faces that belonged on pit bulls—bears that would have towered over the biggest of the present-day grizzlies. Then there were the packs of dire wolves that crushed the bones of their prey with viselike jaws.

When the scavenging birds started boiling down, they must have blackened the sky. Ravens and small vultures would find the carcass first, drawn by the smell of the kill. Some would circle just beyond the reach of the cats in a way that might have signaled bigger birds. Other small vultures would be gathering near the wounds that would attract the big birds, hoping to grab what bits of meat were thrown clear of the melee.

Soon, much bigger scavenging birds would start converging on the carcass, having smelled it or seen the smaller vultures. It's possible that hundreds of birds would be circling the carcass. Eagles with eight- and nine-foot wingspans might dive straight into the crowd, tearing at the choicest bits with their sharp hooked beaks and long, curved talons. When the hides were split, the rest of the birds began moving into position. Storks with extralong beaks started digging into the deepest wounds; other birds raced in from the sides to steal meat out of beaks. Some of the birds might have crawled all the way into the carcass. Condors would have vacuumed their meals out of wounds, opening them wider in the process.

It would have looked like a free-for-all to most of us, but in truth, it probably wasn't. Over the millennia, the birds in the condor's scavenging guild evolved in ways that led them to open spots on carcasses—places inaccessible to other kinds of birds.

The condors would have eaten until they were so groggy that any attempt at a running takeoff would likely end in their tripping and falling flat on their faces. You might have seen a few of them trying it anyway, running and flapping their wings, but when they couldn't get up high enough to find some wind, they would have landed and started again.

When the mammoths and the sloths and the saber-toothed cats disappeared ten thousand years ago, the scavenging guild went with them. The only giant left, the condor, may have been declining ever since. For instance, the evidence Emslie found in the caves of the Grand Canyon seems to show the bird wasn't there for long.

"I found the partial skeletons of three different condor chicks in one of these caves," Emslie said, "and all around these skeletons we found condor eggshell fragments. This was a cave that had been used by the birds on at least an intermittent basis for several hundred years, and as we continued to look around, we found the things the birds were eating. There was a chip of a mastodon tooth in one of the pack-rat middens, and a bone shard from the humerus of an extinct bison. Clearly these animals couldn't have climbed up into caves like these. Condors must have flown them up in pieces for their young."

Emslie's lead climber, Larry Coats, says what may have been the most dramatic find came near the end of the two-year expedition. He and Emslie had just scrambled into a place called Stevens' Cave, named for the climber who left it in a panic after something made a moaning noise and blew his torch out. Coats found the source of that moan when he crawled into a tunnel at the back of the cave—it dumped him out into a secondary cavern that was full of fossilized goat skulls. Coats whirled around when he heard the moan and felt a blast of air—both were coming from a narrow crack that extended out through the cave face. When the wind outside blew a certain way, the moan rose again.

Shortly after that, the lamp attached to the helmet Coats was wearing passed over something oddly white and oblong. It was the skull of a very large bird, half buried in the dirt. Coats yelled for Emslie.

"He bent down to look at the skull, which was partially buried," Coats said. "After a few seconds he yelled out, 'It's a goose!' Then he picked it up and blew off the dust."

It was a perfect condor skull. Emslie whooped with joy. "It was amazingly well preserved," Coats said. "There were bits of tissue hanging off the jaw, and it was completely intact. It looked like the skull of a bird that had died ten or twenty years ago, but when we got the radio-carbon test results back, they indicated it was roughly twelve thousand five hundred years old."

Every single condor bone that Emslie found in the caves turned out to be at least ten thousand years old. This was potentially bad news for the people who had hoped to put condors back into the canyon, since it wasn't likely that a ten-thousand-year-old nest would qualify as proof of "recent" residency. The backers of the condor restoration plan were deeply disappointed, and Emslie didn't know what to tell them.

"What was I supposed to do," Emslie said to me. "Lie about my findings?"

Emslie says his work in the caves helped show what caused the condor's range to collapse: the hunters who marched across the Bering landmass ten thousand years ago. Emslie thinks these early hunters caused an animal apocalypse by methodically killing off most of the continent's giant mammals.

In professional circles, this scenario is known as the Pleistocene blitzkrieg hypothesis, which means it moved like lightning and left little but wreckage in its wake. Next to the theory that the dinosaurs were wiped out when a giant meteor crashed into Earth, it's easily the most controversial theory in the history of extinction studies.

Emslie didn't buy this theory when he first heard it. But when he found the bones in the caves, he changed his mind. The bits of mastodon and sloth he found showed that condors needed giant herbivores, at least in the Grand Canyon; the speed with which the condors vanished showed that early hunters wiped out the giant herbivores. Nothing else could move so quickly and selectively, he says.

"Humans are the only real problem these birds have ever had," he told me. "And that's important. People who think the condor is declining naturally might find it easier to just let the birds go. 'Death with dignity'—isn't that the phrase? I think that's ridiculous. If this species really does have one wing in the grave, it's because we jammed it down there."

It's a wonder they're still with us, I said. By the way, why *is* that? How did condors manage to survive the change that killed the other giant birds?

Go to California, Emslie said. So I did.

———

October 19, 2001—10:47 A.M.: I am stuck in off-peak traffic on the Golden State Freeway near an outlet mall disguised as an Egyptian temple. As a native of this area, I think this is wrong: those walls were supposed to hide a tire factory.

Sitting in my rental car, I think about a scientist I interviewed once, ten or fifteen years ago. His name was John Heyning, and when we met, he drove a modified flatbed truck he called the whalemobile. When dead whales washed up on the region's beaches, Heyning cleared away the crowds and hauled the carcasses off to a warehouse in east Los Angeles. He used to joke that this was the only city on Earth where you could drive a truck with a whale on the back of it and not have anybody notice. He used to say he wanted to pull into a McDonald's and ask for a bun.

And here's the funny part: if a California condor were to soar above the city of Los Angeles today, a dead cetacean on the back of a truck might be the *only* thing it recognized. Whales are probably the things that saved these birds when the era of the supersize herbivores ended. Carcasses the size of mastodons washed up on the beaches all the time back then. They may have been all the condors needed.

"It can't be proved," said Heyning when I called him on my cell phone. "But, I think it may be true." Ten thousand years ago, there were a lot of right whales swimming close to the Pacific Coast, and a lot of gray whales as well. Fur seals, harbor seals, sea lions, and other marine mammals were abundant, and there were many more sharks.

"Condors couldn't have gotten through the hides of the whale on their own," he continued. "On the other hand, they wouldn't have had to." Short-faced bears and eagles would have raced the condors to the beach when the carcass of a whale washed up, ripping holes so big that the careless might have slipped and disappeared into the mounds of blubber. After the Pleistocene, grizzly bears came down in the night to make those openings.

Centuries after that, Los Angeles is doing its best to make it look as if none of these events could have happened. This is a city always pretending to be something other than what it used to be. From the Pacific to the mountains, every bit of the landscape has been terraformed or paved and repaved. The rich, slow river that used to wander down through the center of the L.A. Basin has been straightened and lined with concrete. Flora and fauna were long ago replaced by gangs and movie crews.

The traffic breaks. Not long afterward, I pull into a parking lot of the George C. Page Museum, home of the largest and most spectacular collection of Pleistocene fossils in the world. These bones were pulled from asphalt sumps like Pit 91, just a few steps outside the back door.

The secrets these sumps held were discovered at the end of the nineteenth century, when local cattlemen pulled what they thought were some very large cow bones out of one of them: bones with fifteen-foot-long curling tusks, or giant fangs.

This is where Steve Emslie's professional hero made his intellectual fortune. In the early 1900s, Miller wrote what's still the most eloquent ode to the workings of the tar pits. It starts when a falcon swoops down at a mouse trapped by the sump: "Let the wingtip or a gasping talon break that deceptive surface," Miller wrote, "and the hunter is caught and sooner or later sinks with his quarry. The mouse is no more surely held than the mastodon or the ground sloth. The falcon is no more strongly attached than the sabertooth or the great American lion. Eventually there spread far downwind an odor that was attractive to [another species of bird]; he would swing hypnotically into the wind and to his own undoing."

I walk into the George C. Page Museum to see the skeletons in the glass cabinets: giant eagles, giant storks, and giant vultures. All but the condor have been gone for thousands of years, and at times the condor has seemed ready to follow. Few seem more convinced that it would happen soon than the great Loye Miller, who argued that the principal threat to condors was the passage of time.

"Is not the California condor a senile species that is far past its prime? It was widely distributed and numerically abundant in Pleistocene times (in Florida, Mexico, Texas, New Mexico, Nevada, California) but is now restricted to one or two localities and a numerable population of individuals within the Californias. Is not the condor a species with one foot and even one wing in the grave?"[2]

Those words have been called a noose around the condor's neck, and in a way it's true. This is not a species that's grown old and feeble—that's scientific mumbo jumbo—but it is a creature that evolved to fit a world that's disappeared. The condor is a relic of the Pleistocene epoch, not quite suited to the present day and age.

But does this mean we ought to let the condor fade away? Hell no. I think it means we should do everything possible to keep the condor around. If you see one soaring, think of saber-toothed cats and giant mastodons. Then ask yourself this: "How much would I pay to get some of those animals back?"

three

MORE LIKE RELATIVES

═══

The molokbe (or shaman) pushes his legs through holes in the
stretched skin where the bird's legs had been and laces the skin up
the front of his body. The great wings are tied to his arms. . . . The
molokbe dances slowly. [He] raises his wings. Every few moments
he makes a hissing noise, imitating the condor.

—Dick Smith, *Condor Journal: The
History, Mythology and Reality of the
California Condor* (Capra Press, 1978)

Condors were a thousand times more important to the Indians than they are to you or me. This fact is underlined when anthropologists find the bones of condors in the graves of Indians, and when kids dig up ancient headdresses made of condor feathers out of the corners of caves. In California's San Rafael Mountains, there's an Indian painting of condors on the wall of a particular cave—when the sun rises on the day of the winter solstice every year, the first light that enters the cave lands just below that figure of the bird.

Condors also play a crucial role in the stories told by tribal elders. But the roles they play vary wildly as you move across the state. Surviving members of the Wiyot tribe in Humboldt County, California, say it was the condor that created the present version of the human race, for example. They say Condor and his sister took the job when Above Old Man decided he didn't like the people *he'd* created, and then resolved to kill them with a flood.[1]

This flooding of the world was supposed to be a secret, but Condor managed to find out. According to the Wiyot storytellers, Condor and his sister prepared for the flood by weaving a large, deep covered basket. When the water rose, they got into the basket and sealed it up tight. When the basket stopped rocking and floating around, they opened it and got out. Condor's sister became his wife, and when they bred, there were no eggs. Well-spoken, furless human babies were born instead, and Above Old Man was pleased.

A white ethnographer recorded that story after speaking with the remnants of the Wiyot tribe, which was all but erased on February 26, 1860. That's the day three groups of Yankee settlers armed with hatchets, guns, and knives attacked three different Wiyot settlements, killing whole Wiyot families as they slept. Shortly after that awful day, the U.S. Army herded the remaining Wiyot people through the gates of Fort Humboldt for their own protection.

Members of the Wiyot tribe hold candlelight vigils every year on February 26. Stories about condors that survive catastrophes must have a special resonance to them.

But Condor is also a destroyer of worlds, according to parts of the Mono tribe of Madera County, California. Their tale told that the Condor known to the Mono had a habit of scooping up people and carrying them back to a spring, where Condor would cut off their heads and drain their blood into the water. When the spring was full of blood, Condor always dug a ditch that led straight to Ground Squirrel's house, which he always hoped to flood. Several ground squirrels would then try to flee, but Condor would catch them and fly them to the spring, where he would put them down and pause to take a drink of bloody water. While Condor is drinking, his daughter urges one of the ground squirrels to cut off her father's head, which the ground squirrel manages to do. But when Condor's head comes off, the bloody water he's been drinking "runs forth in every direction," drowning the entire world.

Yet another kind of condor—call it version 3.0—tried to exterminate the human race but got his comeuppance in the end. Members of the Gashowu Yokuts tribe of south-central California said Condor was not only the love child of Coyote and Hawk, but was also a human gambler who shot an arrow at Owl, a powerful magician. Owl responded by causing big black feathers to sprout out of the body of the gambling man, who eventually turned into Condor and flew away. Condor, the former gambler, lived above the Earth, but sometimes he flew down to eat people. Then one day he flew home with three live children, two boys and a girl. At first he told his mother he intended to keep the children as his pets, but then he seemed to change his mind, telling his mother to fatten up the children so that he could eat them for dinner.

Condor's mother started to worry then: When Condor was done eating all the humans, would she herself be the next meal? She decided to get rid of her son and told the children to shoot arrows at him. They did, for half a day, yet Condor didn't even notice. So Mother cut a hole in her son and they all climbed inside. That time, he felt something. Condor flew up and then crashed and died, and his mother and the children burned him. But his eyes burst out of his head and were lost in the brush. Had they been able to find those eyes and burn them in the fire, there would be no condors in the world.

Of all the animals in all the stories told by California's native tribes, only the condor follows such varied story lines. Coyotes were almost always tricksters, for example; grizzly bears, once or future people. But when a giant vulture flies into the stories, you never know what's going to happen.

Another example of the mythological condor's many-splendorness comes from one of the many Yokut villages scattered through what's known as the San Joaquin Valley. This Condor not only tried to kidnap Prairie Falcon's wife but also tried to steal the job of chief from Eagle.

Storytellers from the Chumash tribe of Southern California said the condor started out as an all-white bird that turned black when it flew too close to a fire. Meanwhile, the condor of the Northfork Meno was picking up sleeping people and taking them to a "sky-land," never to return. The wings of the Yokut condor caused eclipses, and sometimes he ate the moon.

This is why the condor is the perfect symbol for the state's divergent Native American tribes. Roughly thirty languages diverged here into more than 130 dialects, so that in most parts of California all you had to do to reach a different language group was walk fifty-one miles. Communities that held more than five hundred people were considered exceptional. Communities of more than a thousand Indians were almost unheard of. Where there was game, these Native Americans hunted; where there were fish, they fished. When wild meat was scarce, they preserved their food, smoking and storing it in big communal warehouses. Loosely affiliated Native Americans rooted themselves in hundreds of "microclimates," usually finding peaceful ways to make the best use of them; in 1769, when Father Junipero Serra toured the state, three hundred thousand Indians were living there in relative harmony. They built no pyramids, formed no empires, and never had tribal councils. They never worshipped Tlaloc or Quetzalcoatl. They failed to invent the wheel.

Instead, they thrived in ways that tended to preserve the richness of the various worlds they occupied. The Native American tribes did burn off the forests, and they sometimes overfished and overhunted. But they didn't do it anywhere near as often as the white settlers did, and by most accounts the damage was not lasting. It's possible that over several thousand years, only one species was put in jeopardy by the California Indians: that was the California condor.

. Death by veneration was a common fate in the era of the Native Americans. There was no doubt the word "death" had a different meaning for them. Most of the state's Indian tribes thought the world consisted of three levels, with the spirit world on top and the physical Earth in the middle. Below the Earth lived powerful and typically malevolent creatures that came out at night. And to the Indians, *all* three of these worlds were real. Animals were always moving back and forth among them, as were special people called shamans, who knew how to borrow powers from the animals, making it possible to travel to the upper and lower worlds.

Borrowing usually meant killing the condor in ritualistic ways, making ceremonial clothing out of its feathers and skin. Shamans danced under sacred wings and feathered capes that sometimes reached the ground. By some accounts, only shamans were allowed to touch these magical items. When particular shamans died, the clothes were said to become cursed.

Shrill-sounding whistles made of condor bones were sometimes played while the shamans danced, strutting slowly and bobbing their heads like the real birds do when they are trying to attract a mate. Many tribes believed that the birds they had killed were brought back to life by the dances: somewhere in the mountains, presumably, they simply reappeared.

No one will ever know how many condors died this way. But one camp of condor experts thinks that the birds might have *benefited* from the killing rituals and dances. These scholars say there's scattered evidence that some tribes stopped killing when the condors seemed to be disappearing. Some of these same tribes were said to treat local condors as invaluable community property. It's also possible that when the native tribes burned the six-foot grasses off their hunting lands, they may have helped the condors by making it easier for them to find food.

But the authors of a recent book on condors take a much more pessimistic view. Noel Snyder, a well-known biologist who was once codirector of the California Condor Recovery Program, says crude calculations seem to show that the rituals hurt the condor badly. After dividing the number of square miles in the condor's historic range by the number of miles occupied by the average tribe of Indians, Snyder and coauthor Helen Snyder concluded that as many as seven hundred different Indian tribal groups could have occupied the condor's rangelands. If even a tenth of those tribes killed condors on a regular basis, the impact would have been huge. "We believe," they wrote in 2001, "that earlier use of condors for ritual purposes may have substantially depressed local condor populations, leading to a continuous overall population decline long before the arrival of the Europeans. Prior to such persecution, condors may have been much more abundant than indicated by any of the historical records."[2]

In 1986, as the last free-flying condors were being trapped and taken to zoos, a group of people who identified themselves as members of the Coastal Clan of the Chumash Nation came forward with a list of demands. No one at the U.S. Bureau of Indian Affairs had ever heard of this clan, but that didn't mean it wasn't real: there had been a time when the Chumash culture dominated large parts of Southern and central California. Ethnographers and Indians agreed that condors were extremely important to the Chumash: when "great chiefs" died, great condors flew them to the Land of the Dead. However, condors needed to be killed to make the journey. This released the condor's supernatural powers, allowing it to fly great distances in small amounts of time.

But this was not the condor known to the self-declared members of the Coastal Chumash Clan; theirs was a bird whose demise in the wild would kill the Chumash culture. The leader of the Coastal Clan, a man named John Sespe, demanded that the trapping cease. When it did not, he warned that great disasters would befall them all.

Not long afterward, an angry group of twenty activists and Indians gathered near the Ventura County office of the California Condor Recovery Program, chanting and disrupting traffic, and handing out flyers that denounced the trapping plan as a violation of religious rights guaranteed by the U.S. Constitution and of Native American rights guaranteed by the Treaty of Guadalupe Hidalgo and the American Indian Religious Freedom Act.[3]

One such release read as follows:

The American people are asked to remember that the condors and the Chumash people, as well as other tribes in California, long predated the US government. As such we call upon the peace [leader] of this nation, the President of the United States, to order his executive branch to immediately cease and desist from capturing the California condor.

A lawyer named Sid Flores, speaking for the activists, was meeting with officials all over the state. In Sacramento, Ventura, Santa Barbara, and Los Angeles, Flores asked that the Coastal Chumash be granted a permanent seat on the Condor Recovery Program, be given permission to watch field biologists wait inside the traps, be awarded exemptions from the law that makes it a felony to own a condor feather, and be offered some of the funding in a condor conservation bill that never made it through the state legislature. Flores said his clients did not seek to undo conservation measures that were already in effect, but they did not want to go along with the plan to send the last birds to the zoo. Flores's clients wanted the last three condors taken out of the zoos and released on Santa Catalina Island, a wildlife refuge west of the Ventura coast. Santa Catalina is considered sacred land by many Chumash, including those officially recognized by the federal government. While the Chumash activists Flores spoke for were not members of a recognized clan, the California Native American Heritage Commission seconded their demands.

"When the condors are gone so are the Chumash," said John Sespe of the Coastal Chumash Clan at a meeting called to calm things down. Sespe then declared that the members of the California Condor Recovery team had committed acts of sacrilege by trapping and handling condors without Chumash overseers on hand. He hinted that his group would go to court to halt the program if they didn't get satisfaction. Furthermore, he added, greater forces were at play here. "Condors control the weather," he said. "If you take them away the weather will do terrible things, I'm telling you."

The unofficial Chumash activists would have faced an uphill battle in court, in part because the same laws that protect their religious rights would have required them to prove that their ideas about condors had deep roots. It might not have been enough to say that a dead uncle had told stories about condors that controlled the weather, or about condors that ceased to exist when they left the wild. Chumash scholars called forth by the government could have argued that there were no hints of stories like these in notes taken by the old ethnographers. Environmental activists from groups such as Earth First! *had* spoken of the condor as a "Thunderbird" with the power to affect the weather, and one or two writers had made the comparison back in the early 1970s. But as far as anyone could tell, there was no Chumash condor with the powers of a Thunderbird. Cynics might have noted that this was the kind of condor you'd invent to stop a trapping program.

But the people who ran the Condor Recovery Program at the time knew they had a potentially catastrophic public-relations situation on their hands. The question wasn't whether the fight could be won, but how to make it go away. Although they hadn't said so publicly, the leaders of the recovery team were appalled by the proposal to move three condors to Santa Catalina Island, mostly because they didn't think the birds would have had a chance in hell of surviving there. Condors aren't built to fly on the winds that swirl over the ocean, and Santa Catalina wasn't big enough to serve as anything other than a glorified outdoor cage. The team members weren't much happier about the proposal to allow a "medicine elder" to perform a special ceremony at the site of every condor capture, as there was always a chance that holding a bird in one- hundred-degree heat for the extra half hour needed to perform the ceremony might seriously injure it. Representatives of these groups were told they couldn't do a special dance at the point of capture, but they were welcome to follow the trapping teams on any federal lands and on private lands if owners gave their permission. Unfortunately, the last of the condors was caught on a ranch whose owners refused to give the Indians permission to go on the land. For several days the Indians waited in a car near the entrance to the ranch, and when they finally abandoned their post, they handed an employee of the U.S. Fish and Wildlife Service a special ceremonial powder, tasking him to sprinkle it on the bird if it was captured.

Pete Bloom caught Igor the next morning. The ceremonial powder provided by the tribal leaders was sprinkled onto Igor's back at the San Diego Wild Animal Park.

four

SWAY OF KINGDOMS

Know ye that on the right hand of the Indies there is an island called California, very near Terrestrial Paradise because of the great ruggedness of the country and the innumerable wild beasts that lived in it, there were many griffins, such as were found in no other parts of the world.

—Garci Rodriguez de Montalvo, *Las Sergas de Esplandián*, 1510

Off the coast of Terrestrial Paradise, 1602: A sailor in the rigging of a Spanish ship would have been the first to see the creatures with the huge black wings rip into the beached carcass of the whale. It's not likely that the man in the rigging would have known what he was looking at. The captain of the ship called the *Santo Tomas* kept his distance from the uncharted shore. Also, it's likely that the sailor in the rigging would have been unable to see much of anything clearly. In the early 1600s, crews on sailing ships like this one were routinely hammered by scurvy, a horrible disease that at the time had no known cure. It was not unusual to lose half a crew to the outbreak, which meant that crewmen who were not about to fall over dead would have to keep on working, ignoring the running sores, and the swollen gums, and the hallucinations.[1]

Our hypothetical crewman would have been illiterate, but he would have known the plot of *Las Sergas de Esplandián*, the wildly adventurous epic poem written by Garci Rodriguez de Montalvo.

He would have known of the warrior women living on de Montalvo's rugged island: "Amazonians" who seduced the crews of passing ships to get pregnant. Afterward, the men were killed and fed to the "innumerable wild beasts," including the griffins with giant wings and the heads and bodies of lions. Male babies born to warrior women were left in the caves of the griffins, who ripped the babies up and fed the pieces to their young.

Picture the confusion on the face of our half-crazed sailor as he squints toward the scene onshore. Could the stories about the flying lions be true? The wings he sees look big enough, but what's that above the shoulders? Could it be the head of a lion?

In the 1600s, it was plausible that scurvy-ridden Spaniards in the rigging could have had these thoughts. These men had been raised to vanquish armies led by warrior eagle gods and giant feathered snakes. Many dreamed of looting cities of gold, and some still feared that the boat they were on would sail off the edge of the Earth. In that context, griffins would have come as no surprise.

But here's what we know for certain: A ship called the *Santo Tomas* really did pass the carcass of a whale being eaten by condors in 1602; sailors, armored soldiers, and a Catholic priest went ashore to investigate the scene. Instead of Amazonians, they found hordes of grizzly bears; instead of griffins, they found condors. That much is recorded in the diaries kept by Father Antonio de la Ascension, the barefoot Carmelite friar and record keeper for the Spanish expedition.

"Birds the shape of turkeys," the friar wrote. "The largest I saw on this voyage." This clipped account is the first known written reference to a California condor, and one of the few firsthand accounts of condors eating dead whales. "Here indeed is material with which to stir the most dormant imagination," wrote condor historian Harry Harris in "The Annals of *Gymnogyps* to 1900."

Civilized man for the first time beholding the greatest Volant bird in human history, and not merely an isolated individual or two, but an immense swarm rending at their food, shuffling about in crowds for a place at the gorge, fighting and slapping with their great wings at their fellows, pushing, tugging at red meat, silently making a great commotion, and in the end stalking drunkenly to a distance with crop too heavy to carry aloft, leaving space for others in the circling throng to descend to the feast.

Father de la Ascension was not so overwhelmed. Like the Spanish scribes who followed him, he rarely did much more than note the bare existence of the condors. I suspect that this was true in part because the word "extinction" had no meaning then, or at least not the meaning it has today. Animals that disappeared were meant to disappear, and it wasn't necessarily permanent; if God wanted something back, all he had to do was snap his fingers.

But the notion that the world was limitless was changing in the 1600s, thanks to a wave of commercial extinction driven by a global fur trade. In the 1500s, European royal families started wearing garments made of fur. By the time our friar saw those condors on the beach, squirrels, foxes, martens, and weasels had been all but erased from the forests of Europe and Siberia.

By the middle of the 1700s, Russian hunters were headed for the condor's feeding grounds. At the far edge of the ocean they paused to kill all Russian coastal fur seals. Then they moved west across the Bering Strait, looking for the rookeries. When the islands in the Bering Sea were bare, the Russians worked their way down the coast of Alaska; when those animals disappeared, the Russians trapped their way down the coast of North America, wiping out the fur seals, harp seals, harbor seals, and walrus.

Condors had less to eat when the Russian traders moved on. Whales were still around, but the beaches weren't thick with them. The carcasses of animals such as tule elk and pronghorn were also available, but that might not have been enough to hold the condors in the long run.

The king of Spain stepped in and saved the species at this point. It's likely that he'd never even heard of condors, but the birds came with the land, and he wanted the land. Rumors that the Russians were preparing forts to protect their pelt-trading interests were enough to convince Juan Carlos III to make a bold move of his own: he would string a line of forts and missions north from the tip of Baja California. Spaniards living in Mexico were ordered to go north and settle in uncharted wilds. Indians would tend to the needs of the settlers after being brought to Jesus. Those who declined would be dealt with. Those who worshipped old gods would be sent to hell.

In 1769, the "Sacred Expedition" left the bustling tip of Baja California, bound for Alta California and the heart of the condor's domain.

Mounted soldiers wearing heavy leather armor rode their horses through the heat, armed with leather shields, heavy guns, heavy swords, and extremely heavy lances. Two other groups of soliders boarded warships in La Paz; the plan was to meet in San Diego and march forward en masse.

Amazing things would happen to members of the Sacred Expedition. Horrible things would happen, too, but I'll get to those stories later. First I want to tell you about the mangy, nervous animals that trailed behind the armored, sweating horsemen; these are the animals that saved the California condor from extinction by dominating Alta California for the next one hundred years at least.

Alta California was God's gift to the animals we call cows; almost everything about the place seemed tailored to their needs. The grazing lands were endless, the weather was perfect, and the relevant diseases were mild. "The growth and development of the range livestock industry in the New World was a phenomenon without precedent," wrote the cattle expert L. T. Burcham in the book *California Range Land*. It was the foundation of the domestic economy of Spanish California.[2]

Cows have been described as "the forward elements in the column of civilization," but in California they were more than that. The line of missions laid out by Father Serra in the wake of this forced march to the north would not have survived for as long as it did but for these scrawny cows. And the condor would be extinct.

These were not the kinds of animals a living soul would recognize as cows: they were not fat and square and uniformly healthy-looking and they did not live their lives in pens. They moved through landscapes never seen by Europeans, mowing down all the native grasses they could find. Most of these cows had long, skinny legs joined to narrow hips and badly swayed backs. Their heads have been described as "combatively coarse," which probably means they were ugly.

But if twitchiness and paranoia can be taken as a sign of animal intelligence, Spanish cows were very smart indeed. "These cattle had a quick, alert restless manner," wrote an early student of the breed. "They have been likened to wild animals, continually sniffing the air for danger."

There was plenty to sniff for. Mountain lions stalked the cows that strayed from unfenced herds, waiting for the chance to leap onto their backs and rip their throats open. Lots of other wildcats tried to do the same.

Then there were the grizzly bears. They were better at killing cows than all the other nonhuman predators combined, and of the humans, only Spanish horsemen armed with rope knives matched the grizzly's lethal speed and grace. Tracy Storer and Lloyd Tevis, authors of *The California Grizzly*, collected several written reports of grizzlies luring cattle in for the kill by lying on their backs in pastures and kicking their paws up in the air. When cows saw this playful scene, they came over to take a close look, and the grizzlies quickly knocked them dead. "The cattle will surround the bear in a wondering and gaping circle," wrote a man who claimed to have seen such killings. "Until [the bear] who is all the while laughing in his paw at their simplicity seizes upon the first cow that comes within the grasp of his terrible claws." Afterward, the grizzly bear walks off with his next meal, "who thus pays the expense of the performance."[3]

Teams of grizzlies may have worked together back then, with one diverting the attention of a mark away from an approaching pair of killers. One barely believable account described a steer that stopped to watch a bear roll himself up into a ball and tumble down a hill into a pasture. "Suddenly at angles from either side two other bears rushed forth," the story goes, "and almost before one could tell what was happening the larger of the two had reached the great steer." At that point, it was over for the unlucky bovine: "Bruin with paw as heavy as lead felled the steer to the earth."

Condors tracked the movements of these bears, just as they once may have tracked the movements of dire wolves and saber-toothed cats. They would have seen what the grizzlies were up to when they rolled around playfully in the pastures, and they would have seen the traps get sprung on the unsuspecting bovines. There's truth to the stories you hear sometimes about vultures that really do swirl over the heads of dying animals. I've been told that condors did it all the time. They know how to wait for the right moment.

Actually, the surveillance may well have been mutual: California grizzlies may well have tracked the movements of the condors that were tracking them. Storer and Tevis couldn't prove that point, but they could offer expert speculation.

The burying of carcasses, a practice of both the grizzly and the mountain lion, may have been an effort to circumvent the big scavenger birds. Conversely, the condor may have aided the grizzly. Soaring aloft at a great height and for much of the daylight hours, one of the birds could locate a carcass. As it dropped down other birds would converge toward the site. Maybe the vulture congregations guided the bear to such a feast. When he arrived, however, the birds would flap their wings vigorously and take off to avoid being included in the bear's repast.

Given the level of cleverness shown by both the bears and the condors, I can't help but wonder whether the big birds ever did more than flag the dying and the dead. Condors stay alive by anticipating sneak attacks, both on themselves and on the animals they'd eat if only they were dead. What I wonder is whether the birds enabled the bears by circling the weak and the dying like the vultures in Western movies do. And if they did these sorts of things, where did they draw the line? When a soaring condor saw a healthy cow get separated from the herd, did it ever pause or circle in ways that might have drawn the grizzlies in?

There's another key link between the condors and the bears: both species tended to thrive near the low, nasty forests known as the chaparral. In the summer, lightning strikes and tiny floating embers triggered instant infernos there, but when the smoke cleared, the forests always grew back thicker than they were before. Grizzly bears and condors seemed to use these forests as a defensive shield, and for a time it served them well.[4]

Old-time nature writers tended to wax eloquent on the subject of chaparral: in the 1920s, these areas were sometimes referred to as nothing less than "Elfin Forests." A writer named Francis Fultz was the one who coined that title, and he does not appear to have been joking; Fultz said he'd take chaparral trees over redwoods any day. "Dame Nature knew her business when she developed the chaparral," he opined. "Without it the mountains would be stark pinnacles and naked ridges, the foothills barren, the rocky slopes and the valleys nothing but beds of cobblestones and gravel." Fultz was among the first to see how varied these forests were, claiming that his "elfin woods" contained a wider range of plants and trees than any other forest in the country.

There's no doubt that Fultz is right about all this: chaparral forests are diverse and they can be magical. If you look around you'll find tiny twisted oak trees and tiny twisted pines, mixed together so tightly that you can't see through them. The smell of mint is everywhere, and in the spring the wildflowers are amazing. You want to crawl around in there until you find the dancing elves.

But when you do, your skin is ripped by knifelike thorns or covered with painful red rashes. Your machete bouces backward off a springy branch and hits you right between the eyes. You creep through a maze of tunnels underneath the chaparral that appears to have sealed it off. Hours later, after crawling through a drainpipe-size opening into the blessed sunlight, you pull your face up out of the dirt when you hear the sound of an approaching bear. Or maybe it's a coyote, or a mountain lion. All you know for sure is that it's big and you don't want to know what it is, so you dive back into the elfin forest, looking for another way out. You trip and plant your face in a hornet's nest or another thorny bush. On the way out, bleeding and screaming, you fail to see the rattlesnake in your path.

Grizzlies used the chaparral forests as a fortified base of operations. Twisting tunnels covered by canopies of the ten-foot trees connected the dens to the pasturelands and to other dens. Sleeping quarters might have multiple entrances. Some tunnels stopped abruptly. Families of bears used these pathways over and over, to the point that in a few locations, generations of grizzly bears literally walked in the footsteps of their predecessors, creating a succession of paw pits that were sometimes eight to ten inches deep and fourteen to sixteen inches apart. When the grizzlies were destroyed by hunters and ranchers spreading all manner of poisons, black bears moved in to further deepen the age-old ruts. I only know of two people who crawled around in the chaparral tunnels because they liked doing it. One is the late Joseph P. Grinnell, the legendary western ecologist who founded the Museum of Vertebrate Zoology at the University of California at Berkeley. The other is a recluse/artist named Jon Schmitt, who used to explore the tunnels when he was observing condor nests as a volunteer for the U.S. Fish and Wildlife Service. Schmitt is exactly the kind of person you'd expect to be living in there—not too tall, big black beard, extremely soft-spoken, a taxidermist and line artist who draws the birds he sees in the wild. Born at least one hundred years too late, he says, and his friends all agree with the assessment.

Schmitt crawled through the chaparral for days at a time in the 1980s. He never met a bear or a mountain lion, but he did learn to recognize the sound angry hornets make when a bear eats half of their nest. He says he often thought of grizzly bears while crawling in the ruts they'd created—how it must have felt to be so big and move so gracefully, melting in and out of the tiniest gaps in endless walls of brush. Schmitt's reveries sometimes got him lost, he once told me, and being lost in the chaparral has certain rewards. Exiting tunnels surrounded by the densest brush, Schmitt emerged into meadows never mowed down by cattle or even lightly grazed by deer, let alone visited by hikers. The lowest branches on full-size oak trees would rise just a few inches off the ground.

They were little time warps, Schmitt says, recalling one bit of chaparral landscape so beautiful that he didn't even want to walk around. "An untouched meadow of native bunch grasses, tightly hemmed and guarded by the densest of chaparral brush," he said. "I've occasionally seen some nice and even extensive beds of bunch grasses in condor country, but they were tiny fragments compared to what I was looking at."

He followed a black bear trail to the middle of the untouched meadow, wondering whether it had once been used by Indians, grizzlies, dire wolves, and saber-toothed cats. "I feel my sense of wonder building," he said. "I closed my eyes and spread this meadow out across the alien foothills of the entire eastern side of the San Joaquin Valley, and then across the whole of the L.A. Basin."

He thinks of this imaginary landscape when he sees the condors soaring. After all, they were there. They saw the grass get eaten by the cattle and replaced by the weeds, or recolonized and covered up by more chaparral.

———

Before the arrival of the cattle herds, hundreds of thousands of native Californians subsisted on a diet of nuts, berries, roots, leaves, fish, and not much meat. They killed game from time to time, but did not tend to count on it: those meals were for special occasions.

The men who ran the missions threw those nuts and berries into the trash, forcing the Indians to eat what they gave them. Indians who did want to eat from these piles were told to change their minds—it was a real "Let them eat steak" atmosphere. Some natives adapted readily to this new diet. One friar wrote that at the mission he ran, it wasn't long before the "average" Indian was asking for ten pounds of beef a day, and some were asking for more. The friar did not say what the Indians were doing with these wagonloads of meat and fat, but the implication was that they were eating it until they couldn't eat any more.

Richard Henry Dana, the wayward New Englander who wrote *Two Years Before the Mast*, said that mealtime wasn't any different in the galleys of ships in the harbors. "During all this time we lived upon nothing but fresh beef," Dana wrote. "Beefsteaks three times a day, morning, noon and night cut thick and fried in fat with the grease poured over [the top]. Round this we sat attacking it with our jackknives and teeth."

New England was the mecca for shoemakers in the late 1700s and early 1800s, and for most of that time, the demand for the hides from California cattle was as bottomless as the supply. Cobblers and middlemen in the docks in New England bought every hide the sailing ships could carry, and the ships found ways to carry tens of thousands at a time. The vessel on which Dana crewed packed forty thousand dried and flattened hides into its hold, and other ships may have carried twice that. From 1822 to 1846, more than a million cowhides reached Boston every year, along with tons of candle tallow made from boiling cow fat. Naturally, the natives were awarded that job.

But a few got lucky. They were allowed to ride with the legendary Spanish horsemen called vaqueros, who worked endlessly to bring the jittery herds to slaughter. Some of the herds stretched more than a mile across, and the horsemen had earned their reputation. Awestruck bystanders watched them ride like demons, twirling lassos with extreme dexterity and skill. Errant cows were almost always brought down with a single arcing throw of the rope.

Sometimes the vaqueros showboated at rodeos, roping and killing all sorts of animals, including California condors. Usually they threw their lassos around the necks of the birds, which doesn't sound especially sporting. Presumably these condors had been rendered incapable of flight before the event. Presumably the throats of the birds were slit when the performances were over.

Soaring condors would have stayed close to the horsemen working on the unfenced ranchos. Vaqueros sometimes killed and skinned cows on the spot to get their hands on an attractive hide. And when the great clouds of dust started rising, it meant the herds were being driven to the rodeo grounds for what was known as the *matanza*, which in English simply means "the slaughter." Fifty to one hundred of the fattest cows would be killed and cut into pieces, with the meat going to the missions, the fat to the tallow pots over the fires, and the hides to the beaches to be stretched, dried, sorted by color, and inspected for impurities.

What a wretched mess the *matanzas* must have left behind them. Near the missions, carcasses without their hides piled up near the "killing trees" the cattle were tied to and then slaughtered; when those piles got too big, the natives had to drag them off and throw them in ravines. Indians who smelled like they'd gone swimming in cow fat tended the tallow fires, slowly stirring the boiling, smoking messes in the vats. Bears and vultures living on the moon would have been attracted by the stench; condors, lacking a sense of smell, would have followed them down. There were reports of grizzlies that swam across rivers and climbed the walls of the missions to get in on the fun. Some walked straight through the centers of towns to get to the killing fields. The condors would have circled down soon afterward, seen the dust and smoke, and gotten there first. They would have gathered by the hundreds in the trees, waiting for the chance to scour the bones.

Through all of this, for many years, the herds kept growing. By the late 1840s, it was often useful to chase down any stray cattle. There were reports of feral herds that grazed between the boundaries of various missions, or in the great swampy valley to the east of the coastal mountains. Spaniards tended to avoid that part of Alta California: when the hostile Indian tribes weren't after you, malarial mosquitoes were.

The cattle era peaked in the 1860s, having outlived the mission system, the Spanish and Mexican governments, many millions of Indians, and most of the California grizzly bears. The decline that followed isn't over yet—there are still some cattle out there—but the end draws closer every day.

The Sacred Expedition sent north to map out missions and save the souls of Indians was in trouble by the time it reached San Diego Bay. Many of the soldiers on the ship that sailed up from La Paz had contracted scurvy; they would have collapsed under the weight of the leather armor and the heavy weapons. Those that did continue were described as looking "skeletal" by a friar who rode with them. Juan Gaspar de Portolá, a military man, led the ghastly horsemen north again, hoping to end up in Monterey.

The Sacred Expedition rode through what the archivist Harry Harris wrote was "utterly unknown and unexplored territory, every mile of it condor country." At one point they camped in an alluvial plain full of willow trees and Indians of "good character." Father Pedro Fages seemed especially stricken by the beauty of the future L.A. Basin, but then a series of earthquakes hit and the expedition rode on.

Later, on some hills near what is now the town of Watsonville, they came upon a burnt-out village that had apparently just been abandoned. The natives who had lived there apparently torched their homes before departing, for reasons unrevealed to the Spaniards riding through the smoky mess. The only thing that hadn't been burned was the carcass of a young California condor. Father Juan Crespi wrote that the carcass had been mounted on a pole in the center of the ruined village. The condor had been skinned and "stuffed with grass," continued Father Crespi. "It appeared to be a royal eagle" left behind to send a message. In "The Annals of *Gymnogyps* to 1900," Harris guessed that the bird was being "raised and fattened for . . . the most important festival on the calendar": the ritual sacrifice of the immortal bird-god Chininich. "Prematurely doing away with this demigod under the circumstances recounted indicates some connection with their hope of personal safety," Harris wrote. "And their belief in the [ability] of *Gymnogyps* to prevail over the death of his friends."

Perhaps the natives thought the bird-god would protect them from mounted, armored soldiers—men who would eventually attempt to force them to accept the Christian God. Thoughts such as those would soon become offenses punished by death. But this time, apparently, the bird-god held, and the Indians were never found.

five

COLLATERAL DAMAGE

The biggest mistake the condor ever made was not evolving bullet-proof vests.

—Lloyd Kiff

The California gold rush hit the condor's world like a meteor from the East. When it did, the mountains crumbled and the rivers died; afterward, squalid human settlements festered like diseases. While the rush was on, a writer went to see the forests near the town of Placerville, returning to report that nothing remained but hillsides covered with stumps. The local river was also gone, said the same man: the water that had once rolled past the town had now been channeled into a mess of hoses that slithered toward the mining sites.[1]

Condors weren't common in the flatlands to the west of the Sierra Nevada—the place where the flow of the rivers slowed and allowed the gold to settle—but this was a catastrophe in which the side effects were worse than the event itself. When the forests near the mines were gone, for instance, the loggers moved up the western side of the Sierra Nevada with the force of an after-blast. Some of the world's Gothic-looking forests were lost in the process. These are the forests John Muir described as "the clearest path into the universe," places so beautiful they'd become a "synonym for God."[2]

Condors roosted in some of those trees, scanning the horizons and waiting for the wind. Thanks to Muir, there's no need to wonder how it felt when the winds hit. He once climbed a hundred-foot Douglas spruce in the middle of a powerful storm. Holding tight to the branches at the top of this tree, Muir felt as if he'd stepped into a flood.

"Yonder it descends in a rush of water-like ripples and sweeps over the bending pines from hill to hill. Nearer we see . . . leaves now speeding by on level currents, now whirling in eddies, and now escaping over the edges of the whirls, soaring aloft on grand upswelling domes of air. Smooth deep currents, cascades, falls, and swirling eddies, sing around every tree and leaf." Condors read the winds Muir did, except that they had better eyes. They would have left the area before a storm like this one could get close, knowing that its winds are not the ones they like to use. But, if by some strange reason they had chosen to stay, they would have known how to skirt the eddies and find the grand upswellings. How they did it—what they saw in the wind—is almost as much a mystery today as it was when Muir was up that tree.[3]

Condors may also have nested in the trees in these grand forests, raising chicks and watching over fledglings. Whether there were more than a few of them is a matter of speculation, but if towering trees with rotted-out cavities near the top were what was needed, the condors would have had a lot of choices. Condors also needed trees with big dead branches near the top for perching and looking around. These would be the branches struck by lightning or singed by fires that burned in the canopies. Muir once wrote of watching fires burn like that, "flaming across the green branches at a height of perhaps 200 feet, entirely cut off from the ground fires, looking like signal beacons." He saw dozens of these canopy fires burning in a forest at once, with those in the distance glowing like "great stars above the forest roof."[4]

What the condors wanted were the trees that were like condors:

gnarled giants with bald, scarred heads and sweeping views of the world. After the gold rush, these were the trees whose stumps were sometimes used as dance floors; trees that had to be felled and *then* blown up with dynamite before the bits would fit into the sawmills.

The market for the wood of these giant trees was minuscule in 1848. Then three hundred thousand people rushed into the region, changing everything forever. When the first flecks of gold were found at Sutter's mill, the human population of California consisted of ten thousand Mexicans, an unknown number of Indians, and hardly anybody else. Five years later, "anybody else" outnumbered Mexicans ten to one and Indians by a margin no one counted, and plans were being laid to build a railroad line that would bring more Yankees west.

The forty-niners called themselves "Argonauts," after the men who followed Jason past all of those Greek monsters to the Golden Fleece. But by some accounts the miners were the monsters. The writer Joaquin Miller, a contemporary, remembered these men as "hairy, bearded six foot giants, hatless and half dressed . . . who shouted in a savage banter" when they weren't stuffing their faces full of "the eternal beans and bacon and coffee, and coffee and bacon and beans." In his book about the gold rush called *Rush for Riches*, historian J. S. Holliday wrote of "lonely men who sought companionship in tented stand-up bars, grog shops, hotels, and gambling 'palaces.' " And while shortages of other basic goods were sometimes widespread, nobody ever seemed to run out of grog.

In 1853, the quantities of liquor imported and presumably consumed included 20,000 barrels of whiskey, 4,000 barrels of rum, 34,000 baskets of champagne, and 156,000 cases of other wines. Beer came in 24,000 hogsheads, 13,000 barrels, and 23,000 cases, while shipments of brandy arrived variously in 9,000 casks, hogsheads and pipes; 13,000 barrels, 26,000 kegs and 6,000 cases. And there were comparable statistics for "unspecified liquor."[5]

The forty-niners were armed to the teeth when they arrived in California. Holliday compares them to a volunteer army massed at the base of the mountains, bristling with weapons of every possible size and vintage. The weapons of choice included tiny derringers and long knives hidden in boots; antiquated muskets dating back to the American Revolution; shotguns used in countless duck hunts back on the farm; firearms stolen by Indians and then stolen back; cavalry rifles and the Spanish blunderbuss from the Mexican-American War; French wheel-lock target-hunting rifles; Colt model 1849 pocket and dragoon pistols; left-handed versions of the Kentucky rifle used by Daniel Boone; air rifles such as the Hawkins Big 50 rifle, better known as the gun that killed the California grizzly.

One forty-niner described his wagon train as a rolling armory, writing that the coats on all the men "are surrounded with a belt in which are stuck two ten-inch rifle pistols and an eight-inch dirk knife, which with a United States rifle or double-barreled shotgun completes our uniform."

Everything these wagon trains went past might as well have had a bull's-eye painted on it—to pass the time, the future miners shot at rocks, trees, bones, animals, and birds, not to mention Indians and each other.[6]

Gunplay and other forms of mayhem were also common in the mining camps: shootings, stabbings, beatings, lynchings, rapes, and arsons were said to be everyday events, along with what Holliday called a "society cankering rapacity" heightened by the "ill-concealed bulges of pistols tucked into every waistband."

Condors near the mining fields were shot dead for the hell of it or crippled because they were there. When two men working for a mining company found a condor sleeping on a ledge near the American River in 1855, one of them, Alonzo Winship, reportedly climbed up to take a closer look and hit it with a shovel. The condor woke up with a broken wing and what must have been a splitting headache; when it started scrambling down the hill, Winship and his shovel followed. According to an unreliable writer who interviewed Winship and his friend Jesse Millikan, the chase continued until the wounded condor stopped running and fell down.

Winship walked to the bird to pick it up.[7] Big mistake. When the man got close, the condor leapt up and attacked, waving its beak around like a mugger waves a stiletto. To escape the bird, our reporter says, Winship climbed to the top of a boulder, where the condor couldn't reach him. Winship sat on the rock and yelled for help. The condor with the broken wing walked around looking "vengeful."

Winship's friend Millikan ran away. Then he came back with a long wooden pole, and Winship got off the boulder and grabbed the far end. The two men tried to bring the condor to the ground by getting on either side of the bird and pressing the pole down on top of its shoulders and wings. In the end, the bird got stuck and the men pushed it to the ground. One of them then jumped on and wrapped a coat around the condor's head, at which point it surrendered.

━━━━━

Life got even harder for the condors in the 1850s as failed miners rushed toward what seemed to be a second chance to get rich quick: the commercial hunting business. All you had to do was take a wagon full of hunters off into the woods, and before you knew it you'd have lots of meaty carcasses to sell to the overpriced restaurants back in the mining towns or to starving pioneers arriving from the East.[8]

The results were catastrophic. Close to half a million tule elk were shot and sold in just a few years, missing only one small herd. Giant populations of deer met a similar fate, as did populations of pronghorn. Other teams of hunters zeroed in on the birds, taking out wagonloads of grouse, quail, duck, geese, and shorebirds, in the course of an afternoon. One group of market hunters collected twenty thousand murre eggs in two days. Others shot frogs, snakes, and ground squirrels.

Nobody important tried to stop the hunts until it was too late. Usually the hunters put themselves out of business by wiping out their product lines. There were no limits on hunting then, and few were even considered. The loss of all that wildlife was thought to be the price of progress, and everyone except John Muir wanted all the progress they could get.

The market hunters moved out of California, killing and selling tons of wild meat to wagon trains coming in on the overland trails. A story in *The California Grizzly* followed one such group of hunters through eastern Washington in 1853, as they piled their wagons high:

There were five horses packed with buffalo robes, of which we had about thirty-five; next four horses packed with bear skins, and several large bear skulls; then two packed with deer skins; two with antelope skins; one with fox and other small skins; seven with dried meat for the use of animals on the journey . . . one with boxes containing the young bear cubs last caught; two with boxes containing wolves, untamed; a mule with foxes and fishers in baskets; and a mule with tools, blankets and camp luggage. Almost all the horses beside the seven specially devoted to the purpose carried more or less dried meat—even those we rode. But the most remarkable portion of the train consisted of the animals which we drove along in a small herd; these were six bears, four wolves, four deer, four antelopes, two Elks and an Indian dog.[9]

This hunting crew was led by Grizzly Adams, a self-proclaimed friend and famed hunter of grizzly bears, and soon to be owner of one of the country's best-known animal menageries. Adams usually hunted for the sport of it, but this time out he hoped to make money selling meat to the wagons on the Oregon Trail. Naturally, the Grizzly One had a rude encounter with a condor on this trip. But before that happened he more than justified his reputation as a man who rarely failed to get his beast. Many years later in the basement of a building in San Francisco, he told the story of the expedition to the writer T. H. Hittell.

Hittell begins the story as Adams and a partner, identified only as a Mr. Gray, lead a heavily armed team of "lost and wandering souls" across the dry side of the Sierra Nevada, stopping now and then to send a line of hunters stomping through the trees. Animals that tried to flee were shot either by the hunters behind them or by the hunters hiding out in front.

After a few days, Adams said, the hunters were followed by a wide variety of predators and scavenging birds, to the point where it was sometimes hard to sleep at night while listening to the "chorusing cries of wolves, coyotes and mountain lions." Then, one night when most of the hunters were asleep, a pack of hungry wolves raced out of the woods at the wagons full of carcasses. Adams told Hittell that he sat up with a rifle and plugged one of the wolves in midstride, stabbing it a couple of times for good measure. The other wolves turned around and ran back into the forest.

Adams shot a mountain lion the very next day, he told Hittell.

After that the team hit a dry spell, marching "over hills and dale and in and out of canyons and arroyos," which were freezing cold and white with snow. When the team made it down below the snow line, Adams immediately gave the order to start hunting again. He told Hittell that the wagons quickly gained a pile of antelope, with Adams dropping a buck. "Supper included roasted mountain-lion meat, which tasted good to the fatigued men."

The hunters also seemed to shoot at everything that moved, picking an eagle at one point. They also shot at crows, magpies, hawks, prairie dogs, grouse, lots of rabbits, and innumerable squirrels. One night, Adams's partner was awakened by scary noises. Adams gave a whistle and heard a bear snort back. At that point the famous hunter moved away from the place where the carcasses were piled. When the beast stepped out into the light of the fire, Adams and his partner both jumped up and shot it through the gut.

The condor was an afterthought. Adams told Hittell he spotted one in a grove of trees near the end of the trip, surrounded by a mob of lesser vultures that would have been keeping their distance while preparing to follow the condor's cues.

"The trees were black with buzzards, which . . . soared and darted down at the camp," wrote Hittell.

> One giant bird was particularly aggressive, as if he were the King of the Vultures. Adams fired at the great bird and broke one of its wings. When he approached the wounded condor, now flopping on the ground, the huge bird made such a show of ferocity that he decided to give its powerful beak a wide berth.

Adams handed his pistol to an Indian and told him to shoot the condor in the head.

Grizzly Adams didn't want to waste his time on condors. When he got old and gave up hunting, he would change that attitude. For years Adams tried to add a California condor to the protozoos he led down the main streets of American cities—lines of big fierce animals collected and "befriended" by the famous hunter himself. The menagerie was anchored by a pair of grizzlies Adams had long kept as pets; at various times the line behind the "loyal" bears included monkeys, wolves, eagles, elk, cougars, and baboons. Adams, marching with the bears, wore a wolf skull as a kind of hat: the pelt of the wolf, still stuck to the animal's skull, spilled halfway down Adams's back. City people were supposed to follow these parades to a building with more animals inside it. That's where old man Adams made his living selling tickets.

There are no records of a California condor in these lines. Apparently, the only condor Adams ever saw was the one he told his Indian to shoot. At one point, he bought an Andean condor. It wasn't the same.

———

California condors got meaner and bigger in the late 1800s, assuming everything you read in the papers and magazines was true. Wingspans that had never stretched for more than ten feet grew to fourteen feet and more. Condors that had eaten only dead things began swooping down on bleating lambs and family pets. The talons on the feet of the condors seemed to grow much longer and sharper. In 1858, *Hutchings' California Magazine* published a drawing of the bird flying off with a helpless rabbit impaled on its big black talons.[10]

This was the condor the people "Back East" expected to find "Out West": a good-for-nothing outlaw and double-crossing thief that deserved to meet its maker. The fact that condors of this kind were fictional didn't seem to matter much at all.

Those who felt the need to explain would have argued (incorrectly) that a bird that makes its living eating rotting flesh has got to be some kind of a public health threat. On top of that, as almost every rancher in the state would tell you with perfect certainty, the condor was a big-league thief. Everybody seemed to know a hunter who'd bagged a great big animal once and then remembered he had pressing business elsewhere, only to return to find that the trophy kill was now a skeleton surrounded by a dozen woozy condors, or two dozen condors, or a hundred or two hundred condors. Ornithologist Adolphus Heerman said that scenario repeated itself frequently while he was conducting a survey of animals found near the likely path of the railways.

> We have often passed several hours without a single one of the species being in sight, but on bringing down any large game ere the body had grown cold these birds might be seen rising above the horizon and slowly creeping towards us, intent on their share of the prey. Nor in the absence of the hunter will his game be exempt from their ravenous appetite, though it be hidden carefully and covered by shrubbery and heavy branches: I have known these marauders to drag forth from its concealment and devour a deer within an hour.[11]
>
> —Pacific Railroad Surveys, as cited in
> "The Annals of *Gymnogyps* to 1900"

The largest convergence of condors on record is alleged to have taken place in 1850 on a ranch in central California, according to a journalist not widely known for accurate reporting, A. S. Taylor. Taylor interviewed a rancher who was taking a wagonload of beef to market one day when his wagon hit a bump that apparently caused the beef to bounce out of the back and onto the ground. Amazingly, the rancher didn't hear the thump or feel the wagon get lighter. When he finally did turn the horses around and return to the scene of the bump, the rancher claimed to have looked up and seen more than three hundred condors hovering over what was left of the fallen meat. Taylor writes that the rancher was amazed by the speed with which these condors appeared out of nowhere, "as if they had dropped out of some cavern in the sky."

Taylor's stories, once widely read, are now known mostly for their whoppers, like the ones that repeated the mistaken claim that the eggs of condors were not the pale bluish-green they actually were, but jet black. All the same, a gathering of more than three hundred condors is at least a possibility, and gatherings of more than two hundred are easy to accept: reliable reports of dozens of condors converging were once common. I know ornithologists who'd trade some of their body parts for the chance to see three hundred condors in one place today.

It was about 1850 that condors started getting sick and dying in great numbers, and getting sick and dying is not something condors tend to do. Over the millennia, for obvious reasons, this species has become amazingly resistant to the microscopic organisms that bloom in rotting carcasses, but now the birds on some dead animals were convulsing until their lungs collapsed; they were dying of suffocation.

This is what happens when you eat a meal laced with huge amounts of strychnine, the animal-killing poison of choice among ranchers in those days. Strychnine was the weapon of mass destruction in the war between the ranchers and grizzlies. That war was raging on all fronts in the late 1800s, with some towns funding "community hunts" and others bringing in hired guns. Many of these hired guns claimed to kill their grizzlies by the dozens, and one man said he killed at least two hundred in the course of a long and fruitful career as a bear exterminator.

In most of the state, the grizzlies were in full retreat when the market-hunting teams came in to get them. Afterward the bears were usually riddled with arrows and pumped full of lead, and some were impaled on spikes at the bottoms of hidden trenches. At times the bears were hunted down and shot en masse by mounted federal troops; at other times they were roped and captured by teams of Spanish horsemen. Guns hidden near carcasses were rigged to fire when the bears started pulling at the meat. Susan Snyder, a writer and archivist at the Bancroft Library at the University of California-Berkeley, lists some of the reasons the grizzlies had to die in a book called *Bear in Mind*:

> Grizzlies were killed for their meat, gallbladders, oil and sometimes for their pelts. They were hunted out of fear and the need to protect property; for sport, for power and for target practice. The annihilation of the California Grizzlies was synonymous with progress, civilization, control, management and commerce. They were killed because they had no respect for property, because the country had to be made safe for beef, because they possessed the inherent capability of doing damage. As adaptable as grizzlies are, their defenses did not increase as new weapons were developed to use against them—strychnine, whaling guns, pendulum traps, liquor-laced bait and other ingenious means of slaughter.[12]

The grizzlies that held out the longest were the ones that lived in places rarely visited by humans. In California, most of those places were mountainous and thick with chaparral. Outside hunters didn't seem to get a lot of local help here; when they thought they were getting help they were sometimes getting scammed. Tales of especially ferocious bears seemed to grow as the species dwindled. In the end, these "ghost bears" came to be a source of perverse local pride.

Some of the nastiest bears around were rumored to live in the mountains that joined the coastal range with the Sierra Nevada, at least that's the story told by the reporter sent down from San Francisco. These were the bears that were said to stalk the treacherous mountain passes, waiting for unsuspecting passersby. Clipped obituaries in local papers such as *The Newhall Signal* often followed:

Mr. Beal was found dead a few days ago in the vicinity of Gorman's near Fort Tejon. His skull was crushed and his body fearfully mangled by the grizzly bear. His Winchester rifle was broken to pieces and the portions scattered about, quite a distance from the body. The gun barrel contained an empty shell.

There's no proof that grizzlies living in the Transverse mountain ranges were either extra big or extra mean: maybe the locals who came to look for them were just better exaggerators. Alan Kelly, a famous correspondent with the *San Francisco Examiner* in the 1880s, began a book by expressing disgust with lies like these—and then he went on to describe the havoc wreaked by a monster grizzly known variously as Old Clubfoot, Old Whitehead, and Old Mutah. According to Kelly, Old Clubfoot lived in the mountains near the town of Piru, where he slit the throats of hundreds of cattle and at least one Mexican herder. When a local hunter followed Old Clubfoot's tracks up Piru Creek one day, the bear "loomed up madder than a hornet" and chased the hunter through the woods. Eventually, the hunter outsmarted Old Clubfoot and filled the grizzly full of lead. Or so wrote Kelly:

When I examined the dead Grizzly I found the most singular thing I ever came across [said the hunter]. In the sole of his right forepaw was an ivory-handled bowie knife, firmly embedded and partly surrounded by calloused gristle and hard as bone . . . evidently he walked on that heel to keep the blade from striking stones and getting dulled.

This is only one of Kelly's many fabrications. Furthermore, it seems the locals really took him for a ride. According to historian Charles Outland, the guides Kelly hired to help him hunt for grizzlies never took him anywhere near the bears, preferring to wait for him to go to sleep, then fake bear prints near the entrance to his tent. In the morning, they would swear that these were the paw prints of Old Clubfoot himself, after which Kelly would be led off in hot pursuit of the monster.

Kelly did return to San Francisco with a grizzly he called Monarch, which was probably the last of the California grizzlies when it died in captivity. He writes about trapping this bear in his book. Outland says he probably bought it.

These tall tales are important because people believed them at the time, and these beliefs helped keep them away from the condor's mountain strongholds. Fears of being eaten by grizzlies in the Transverse Ranges must have helped protect the condors living up above the chaparral, as did the relatively common notion that people in the area were mostly crooks and thieves.

Even the law-abiding citizens seemed to be a little off: Ari Hopper, a local rancher who would figure in the condor's future, is said to have acquired his foghorn voice by accidentally drinking a large container of lye. The roads that wound up and over the peaks might as well have led directly to hell. The worst of these roads was that which zigzagged up and out of the southern end of the Great

Central Valley, passing over newly opened earthquake fissures and under walls of rock that always seemed on the verge of collapse. Near the Tejon Pass the mountains got so steep that the wagons had to be taken apart in order to haul them up and over the summit. The only people who liked this process were the local gangs of bandits. The most famous of these bandits was the Mexican-American antihero Tiburcio Vasquez, who staged a series of daring raids on wealthy Yankees near Los Angeles before spending his last two years robbing wagons near Tejon Pass. When mounted posses tried to hunt down Vasquez, he and his gang disappeared, hiding in a maze of giant boulders now called the Vasquez rocks.[13]

Fears such as those sealed the region's reputation as a kind of forbidden zone. One late-nineteenth-century visitor called the place a "terra incognita," inhabited almost exclusively by "grizzly bears, mountain sheep, California lions, rattlesnakes and other such friendly animals."

The owner of a grocery store called Lechler's in the town of Piru used to hang bits and pieces of this history on his walls, and if you gave Harry Lechler the chance, he could make up all the extra stories you wanted on the spot. My brother, Peter, and I used to hang around in Lechler's in the early sixties, when Peter was in the first grade at Piru Elementary School and I was in the third. I remember lots of dusty miner's tools and a bear head hanging on the wall. Harry Lechler seemed to know every story ever told about those mountains, ranging from tales about the ghost of Old Clubfoot to ones about the true location of the Los Padres gold mine.

Lechler also liked to stress that the first recorded gold strike in California history did not take place at Sutter's mill in 1848, but under an oak tree in a canyon a few miles from his store on March 9, 1842. On that day a man named Jose Lopez dug up some onions and saw gold clinging to the roots. Hundreds of miners rushed up from Mexico and from other parts of California, and some carried their gold dust in the hollowed-out quills of giant feathers plucked from the wings of condors.[14]

At the start of the twentieth century, there may have been as few as twelve surviving California condors, according to a former secretary of the Smithsonian Institution. Alexander Wetmore never explained how he came to that conclusion, and many western ornithologists dismissed his guess. Unfortunately, almost everyone agreed that the condor was headed toward Wetmore's low-end estimate, and many feared the bird was doomed. "There is no doubt that the species is in the process of extinction," wrote ornithologist J. G. Cooper in 1890. "I can testify myself that from my first observation of it in California in 1855, I have seen fewer every year." Cooper thought the reasons for the condor's downward spiral were clear: One was the trend toward smaller grazing lands and larger citrus orchards; the other was "the foolish habit of men and boys" honing their shooting skills by trying to blow the heads off all the condors they saw. In 1890, a state law made it a crime to shoot at condors just because they were there, but it had never been enforced.

Apparently, Cooper hadn't seen the condor for eight years when he wrote those words, not since he'd encountered one on a beach in what is now called Orange County in the spring of 1872: "I approached it, being on foot and not attempting to conceal myself, as I was armed only with a hammer and not prepared to attack the bird." Surprisingly, the condor did not fly away as Cooper walked toward it. Rather it looked at the hammer-toting scientist with its "eyes wide open, as unconcerned as if it considered me a brother biped."[15]

Cooper got closer. The condor stared. Cooper got closer still. "As

I had never succeeded in shooting one of these birds on account of their shyness and because I rarely carried a rifle, shot being nearly useless for killing them, I debated whether or not I should take advantage of this lucky chance and kill it with my hammer."

This is where the double-crossing outlaw condor created by the gold rush would have "savagely" charged at Cooper, but the bird near the beach in Orange County was either sick or bored. It barely moved a muscle when Cooper came within striking distance, "except to open its bill in a lazy way when I pointed the hammer at it." At the end of this weird standoff, Cooper did the other unexpected thing: instead of bringing his hammer down on the condor's head, he decided to leave it alone. "I turned and left it to fulfill its destiny," he said.

But at the end of the nineteenth century there were a lot of naturalists who would have brought the hammer crashing down.

SKIN RECORD

None but a naturalist can appreciate a naturalist's feelings—his delight amounting to ecstasy—when a specimen such as he has never before seen meets his eye.

—John K. Townsend, bird collector,
*Narrative of a Journey Across the Rocky
Mountains to the Columbia River,* 1849

When word got out that the condor was getting scarce, every junk pile dubbed "museum" saw to it that it got its share.

—William H. Dawson

The California condor on plate 426 of Audubon's *The Birds of North America* is clearly up to something. The body's out of balance and there's something oddly hunched about the wings. The bird has whipped its head around to look at us through those bulging eyes. It wants to know who the hell we are and what the hell we're doing in its roost tree.

Birds that look alive and self-aware are the reason John James Audubon is still a household name, I think. Stare at almost any of his plates for a couple of minutes and you'll see what I mean. First you'll start to wonder where the bird you're looking at is planning to go when you put the book away; then you'll want to know what it's *thinking* about, and what it thinks about you.

Audubon's ability to capture live birds inside his masterworks is even more impressive when you think about the kinds of birds he

used as his models. All of them were dead, and most had been gut-
ted, skinned, and sewn back together again, with wires and wads of
padding instead of bone and muscle, and cotton balls pushing out
the eyelids. Mounted birds from good museums featured glass eyes
and lifelike poses. But Audubon also used bullet-riddled birds
whose wounds were covered by rearranged feathers. Taxidermy was
becoming a high art in the 1840s, when Audubon was at work on
The Birds of America. Trophy hunters, artists, and naturalists
mounted birds to study. They were the closest things to photographs available then, and even when photographers began coming around, there was nothing quite so useful as a nicely stuffed
bird.

Audubon never saw a living condor soar on the thermal winds,
and if he ever saw a live one at all, he never mentioned it. It's likely
that the bird he reanimated was in bad shape, even for a carcass: a
drab-looking juvenile with a massive head wound, a shattered
wing, and a severed spinal cord. The most likely source of this condor was a rising star who'd soon become one of Audubon's bitterest
rivals. At the time John Kirk Townsend of Philadelphia collected
this bird in his late twenties, he and the artist were the best of
friends.

When his short, spectacular career began in the early 1800s,
John Kirk Townsend was the very model of a modern natural scientist, which means in part that from the vantage point of the twenty-
first century, some of his achievements seem a little profane.
Townsend was a deeply religious man who robbed the graves of Indians. He was an upper-crust Philadelphian who traveled through
the wilderness with riotous drunks. He was a meticulous scientist
and medical doctor who once jammed his head into the sliced-
open stomach of a bison carcass to quench a terrible thirst. "I
pushed aside the reeking ventricles," he wrote, "drinking until

forced to breathe." Townsend may also be the only ornithologist ever to discover a new species while simultaneously wiping it out: that would be the Townsend's bunting, which was apparently found and erased with a single shot in 1835.[1]

Townsend was a careful correspondent with a good eye for absurdity; he didn't mind describing the ridiculous things he sometimes had to do to get his specimens and to keep them in good shape until they could be hauled back east. He writes of waking up in the morning to find that his guides had drained the jugs of whiskey used as a preservative even though the jugs had lots of small dead birds inside them, floating in the booze.

But in my opinion, the story of how Townsend got his giant vulture is the most gripping story he ever wrote and easily the most sensational true story ever written by a condor collector.

The trip that made the story possible began in 1834, when Townsend and a fellow scientist won a $125 grant that allowed them to join a party of seventy men headed west to the coast of Oregon. Townsend and the famous Harvard botanist Thomas Nuttall spent the next several years collecting plants and animals for the American Philosophical Society and the Academy of Natural Sciences in Philadelphia. In Townsend's words, "Mr. N and I were up before dawn [every day], strolling through the umbrageous forest, inhaling the fresh air and making the echoes ring with the report of guns."

In his *Narrative of a Journey Across the Rocky Mountains to the Columbia River,* Townsend marvels at the things he saw on the trip west: flocks of passenger pigeons that seemed to number in the billions, bison herds that looked as if they could cover the world, and brilliantly colored Carolina parakeets that seemed "entirely unsuspicious" of the dangers posed by passing hunters. But the creature Townsend most wanted to see did not appear to him at first. "The

Great Californian" was the name he used to describe the California condor. "I kept a sharp lookout for this rare and interesting bird in all situations," he wrote. "But not one did I see" while traveling west to the Pacific.

Townsend and Nuttall began the trip in identical outfits they'd bought together in St. Louis: dark leather trousers, green felt coats, and white wool hats with wide brims. They could not possibly have fit in less with the rest of their traveling party, but that may have been the idea. These were men who got drunk when the sun came up and brawled until well into the night. By comparison, the oddly dressed scientists spent almost every day collecting specimens, and almost every night pressing the plants and skinning the animals. By the end of the trip, they'd filled several wagons full of exotic species that most had never seen before on the East Coast.

There weren't any condors in those wagons when the formal expedition ended. Townsend apparently gave up hope of ever seeing the giant vulture and sailed off to the Sandwich Islands (now known as the Hawaiian Islands), where he stayed until 1837. Then, before returning to Philadelphia, Townsend took a trip up the Willamette River to watch millions of salmon swim back to the shallow waters they'd been born in, then spawn and die, and "there, to my inexpressible joy, soared a California condor, seemingly intent upon watching the motions of his puny relatives below."

The condor watched the dying salmon leaping up onto the banks of the Willamette River, according to an article Townsend later wrote for the Linnean Society of the Association of Pennsylvania Colleges. Then, shortly after Townsend arrived, he saw the bird shoot down "like an arrow" toward the riverbank below him, landing on the writhing body of a salmon that had just leapt out of the water.

Thrilled to finally see the bird he'd dreamed of in the flesh, Townsend did what any good naturalist would have done in 1837: he raised his rifle and shot it, shattering a wing and bringing the condor down on the far side of the river, where it landed in what appeared to be a lifeless lump. Seeing no signs of movement, Townsend took off all his clothes and swam to the far side of the river, where he "sprang upon the shore and ran with delighted haste to secure the much coveted and valuable specimen." But the condor was back on its feet by then, looking like it wanted revenge.[2]

Townsend, now completely nude, chased the condor around for a while. Then he saw Indian families coming out of the woods. "The inhabitants of an Indian village near," he wrote, "men, women, children, and dogs, startled by the sound of my gun, were flocking out to see what was the matter." Townsend writes of looking around in vain for a stick with which to hit the giant bird: "None was to be found, and my only weapons were stones, with which I continued, for a considerable time, to pelt the vulture."

The natives stood behind him, amused. "I heard more than once the loud obstreperous laugh of the women," Townsend writes, adding that the women seemed to laugh especially loudly "when the Vulture was flapping after me, and I was throwing sand in his eyes with my naked feet."

Other men—me for instance—would have been less willing to expose their naked bodies to a bird with a razor-sharp beak. But Townsend writes that he chose to stay and keep throwing rocks at the condor even as it "dashed furiously at me, hissing like an angry serpent." After about an hour, one of those rocks smashed into the condor's head, at least knocking it out. Townsend turned and borrowed a hunting knife from one of the Indians, using it to sever the condor's spinal cord.

No one would have questioned Townsend's need to kill that California condor, least of all his colleagues in the scientific world. Curators all over Europe and the United States had been yearning to acquire a condor skin since the late 1700s, when the British Museum put the first one to leave America on display. Bird lovers from all over Europe came to see the dead giant from the far side of the world, despite the fact that the beak of the bird was made out of colored wax. (The real beak may have broken off in transit, or been blown off the condor's head by a bullet: no one knows and no one seemed to care.)[3]

But while many museums and universities made it known that they would pay what it took to get a condor carcass, most of them never got the chance to reach into their wallets. Lewis and Clark shot more than one condor in 1805, but for decades almost everybody else who tried failed even to see the birds. One of the exceptions was David Douglas, a Canadian naturalist who wrote of seeing many condors near Vancouver in 1827. "This magnate of the air was ever hovering around," he writes. "In flight he is the most majestic bird I have ever seen." Douglas was also among the first to marvel at the speed with which the condors found fresh carcasses, writing that "only a few minutes after a horse or another large animal gives up the ghost, they may be descried like specks in aether."

All but one of the condors that were sent overseas left the country as carcasses. The one that didn't sailed out of Monterey in the 1860s in the hold of a merchant ship. It's likely that the bird was kept belowdecks in a big wooden crate, into which the condor would have vomited and shat on a more or less constant basis. The stench would have made it impossible to keep the bird in a passenger's cabin. Someone would have brought the bird meat and water every couple of days, checking to make sure the thing inside the crate was still alive.[4]

That's an image that sticks with me: a seasick condor in a wooden crate in the darkness of a hold for weeks. What did it do when a door swung open and the light flooded in? What did the man who brought the food think he saw inside the crate? Maybe the condor's beak came out between the slats as the man with the food approached. Maybe he heard a fiery hiss or the flapping of cramped wings.

There are no published records of the trip that took this condor south to Panama, or of the means by which the bird was moved through the wet, green jungles to the western shore of the Caribbean Sea. This is where a second ship took possession of the crated bird, moving it across the Atlantic Ocean and up the River Thames, where it's likely that the crate was placed in the back of a horse-drawn wagon and hauled over crowded cobblestone streets to the London Zoo.

The eyes and the lungs of the condor would have been burning by then, since London's skies were thick with the coal smoke pouring from millions of chimneys.

The condor survived the harrowing trip to the London Zoo. If anything did go wrong, it was not thought to be worth including in the minutes of the London Zoological Society.

The general demand for skins that could be put on display in museums rose sharply in the late 1830s, when taxidermists learned that arsenic powder kept these skins from rotting. The taxidermists who used this powder knew it was dangerous stuff, but many thought the risks could be eliminated by mixing in other powders and applying it with care.

It's my guess that Townsend put a lot of arsenic powder on the wagonloads of carcasses sent back to Philadelphia after the western expedition, and on the condor skin that followed. The condition of that first group of skins delighted John James Audubon, who bought ninety-three of them. "Such beauties, such rarities," he told a friend. "And hang me if you do not echo my saying so when you see them!" But Audubon never mentioned Townsend's condor skin, even though he may have used it. By the time it arrived, he was Townsend's enemy, calling the explorer a "lazy" man who'd collected skins of little use.

Townsend had been planning to write a series of his own books about the birds of the United States, with lavish illustrations to be provided by various French artists. But he couldn't compete with Audubon's masterwork. Townsend only published the first book in his planned series before he went broke and junked the project. Townsend's brilliant future began to fade, and he had begun to lose his health. In 1851, after years of sickness, he died. The cause of death by most accounts was arsenic poisoning.

———

Bird collectors hired by museums hunted California's condors on an intermittent basis for most of the eighteenth and part of the nineteenth centuries. What they did more often was complain about how hard it was to find a condor to shoot. Some of these hunters thought the birds were shy and inclined to stay away from people, while others talked about the near-impenetrability of the mountains the condors lived in.

Fears for the future of the species were expressed on a regular basis. Just as regularly, sport hunters or ranchers got the blame. Overzealous private collectors were occasionally added to this list of suspects, along with the "oologists" who bought and sold the condors' eggs.

The only group the scientists did not blame was their own. Suggestions to the contrary were met with indignation at first, and then with explanations of the many things the scientists learned from bird skins. These explanations were crystallized in a 1915 essay written by Joseph P. Grinnell, one of the most important and accomplished natural scientists in his day. "It is with considerable apprehension that I have observed an unmistakable decrease in the number of collectors," he wrote. "Matters of precision and accuracy are no doubt suffering as a consequence of this forsaking of the 'shotgun method.'" In Grinnell's opinion, this change was not only a threat to the future of ornithology but a possible threat to the entire scientific method:

> The training involved in bird collecting can surely be given some credit in several cases of eminent men of science who are now valuable contributors to science in other fields. The making of natural history collections is useful as a developmental factor, even if dropped after a few of the earlier years in a man's career. Collecting develops scientific capacity; it combines outdoor physical exercise with an appropriate proportion of mental effort . . . The ultimate fate of practically all private collections is the college or the museum. Very few bird skins, for instance, are destroyed except through fire or other catastrophe. They live on and on.[5]

Grinnell did not want to see the budding science of ecology turned into a hobby pursued by "dilettantes," "amateurs," and "extreme sentimentalists." Bird protectionists were to be thanked for responding to turn-of-the-century warnings of the damage being done by the market hunters and ladies wearing big feathered hats, but now Grinnell thought the activists went too far when they tried to plug the scientists' guns. Eastern states were then considering plans to ban all forms of bird collecting, lest more species go the way of the extinct passenger pigeon. Grinnell agreed that for the rarest birds, the collectors should be forced to make exceptions: "Limitations may be properly imposed . . . by excepting species like the Ivory-billed woodpecker or the Carolina parakeet."

Every time I read those words I wonder why Grinnell didn't put the condor on his short list of critically endangered bird species, given the widespread hunch that the condor was already over the edge. Grinnell was never shy about speaking his mind, and he had special interest in the condor: in his teens he spent countless hours watching an active nest site he'd discovered near the town of Pasadena. "It's an out-of-the-way corner of the mountains, so I think they are safe," he wrote in a letter to a friend in May 1905. "I and my boys who know of this have agreed to leave these birds strictly alone; so that unless some fool gunner gets a shot at them they will doubtless nest in the same place next year."

Grinnell was a scientific prodigy, famous for his fieldwork and his skin collection by the time he was old enough to drink. Stories have been told about the time he memorized a thick book of scientific terms, and of how he once looked out of the moving buggy he was in and announced that local kangaroo rats had begun their breeding season, even though there wasn't a single kangaroo rat to be seen. ("He could see the impact of their scrotal sacs in the sand," wrote his close friend Alden Miller, the son of the field biologist Loye Miller. "He was soon proved right by the yield of the traplines.") Grinnell grew up to be the editor of a leading ornithological journal, the director of the Museum of Vertebrate Zoology at the University of California at Berkeley, the creator of a famously precise set of rules for scientific note-taking in the field, and the author of landmark studies of everything from grizzly bears to ecological niche theories. For most of his career he was the grand pooh-bah of natural-history studies west of the Missouri.

Yet despite his early interest in the condor, he was quiet on the subject. One explanation is that early on, Grinnell didn't think the birds were as close to extinction as the failed collectors kept saying; he knew ranchers who said they still saw condors by the dozens.

But here's a theory I like better than that: I think Grinnell worked hard to keep the condors out of the limelight, as a way of keeping them away from people.

Grinnell was among the first to see the condor as a creature of the wilderness and not as an outlaw bird that flew off with live prey. Unfortunately, there were lots of humans who wanted to get close to the bird, in order to admire it or shoot it or steal its eggs or simply chase it around. I think Grinnell tried to solve this problem in the simplest possible way: if people forgot the bird was around, they wouldn't go looking for it.

This approach was a gamble for an obvious reason: until the end of the 1930s, no one had ever spent more than a few days at a time watching condors lead their lives. No one had ever tried to count the number of condors left in the world, or to systematically study the impact of the skin collectors on the condor. Grinnell helped solve that problem when he sent a graduate student up into the Santa Susana Mountains to do just that in 1939. But before we get to that, we need to take a closer look at the damage done by two groups of condor killers.

The first group didn't really kill the birds. Instead, they kept them as pets. References to condors kept on leashes in people's front yards started turning up before the gold rush, and they did not stop turning up until almost a century later. In 1846, a man named William Gamble wrote that he had long been impressed by "the great disposition of the Vulture to become domesticated." Gamble said he knew of an Andean condor that roamed the streets of a Peruvian city for years:

It would follow and walk alongside a person like a dog for a considerable distance and offer no resistance to being handled or having its feathers smoothed down. It would ascend a long hill leading to the part of the city where the foreigners resided, and when tired of the place or after having procured all it could obtain to eat, it would spread its large wings and soar down to the city, alighting perhaps on a steeple or some other lofty point . . . I think I have never met with any bird which exhibited more tameness or more confidence in man than this large and remarkable condor.

California's condor owners said their birds could do all that and more. One sad owner wrote that his pet condor was jumping in and out of a child's wagon when it slipped and broke its leg, forcing the owner to shoot it. A condor tied to a leash tied to something else was killed by a dog. A bird that was supposed to be kept as a pet died when it was fatally injured by a lasso. Condors kept as pets and then returned to the wild would have had a hard time surviving. No one's ever tried to add it all up, but we do know the pet-taking habit extended well into the twentieth century, when the world's most famous condor pet came into the picture. That was a bird the photographer and naturalist William Finley took from the secret nest near Pasadena, after being led there by his friend and colleague Grinnell. The condor, known as "the General," was supposed to be the subject of an elaborate photographic documentary, beginning on the day the General hatched from his egg. To take these pictures, Finley dragged an old-fashioned camera up and down the mountain many times. One day, when a parent condor seemed to shun the chick, Finley picked it up and took it home with him to Oregon, where he locked the condor into a long, narrow wire-mesh cage built around the stump of an apple tree. The

General was allowed to roam the yard for a time each afternoon; when Finley was late with meals the bird climbed up the side of the enclosure, sticking his beak through the openings and looking around for his master.[6]

Finley wrote a series of articles about the General for national magazines. In all of them the bird was described as being "gentler" than any cat or dog. "Why should such a creature be revolting?" he asked. "He was not ugly to me. Behind his rough exterior and his appearance of savageness, this young condor showed a nature that was full of love and gentleness." Finley wrote that the General was shy in the presence of men and unusually fearful of "strange women, which we thought was due to their manner of dress."

Finley's bird was a new and different version of the condor: one that was competing with the family dog for the title of man's best friend. You wouldn't want to leave this bird in charge of the kids, but you wouldn't ever think of killing it, especially when it was playfully sticking its beak up your pant legs. "One might think a person could have little attachment for a vulture," Finley wrote. "[But] there is nothing strange or treacherous in the condor nature. General undoubtedly felt a strong love for society. He liked to be petted and amused. He preferred to be near me rather than alone. His intelligence was surprising at times."

Finley, like his close friend Grinnell, thought the condor's fate was inextricably linked to the fate of the condor's wild homelands, which turned his efforts to humanize condors into a double-edged sword. When Finley wrote the first of his many odes to the bird he called the General, he honored a fad that pulled condors off the very land he wanted to preserve. None of these birds would ever breed or lay a fertile egg again. Many were killed when their owners got bored or tired of seeing carcasses in the yard. Some were left to be released when they got too big, even though these birds would by then have been behaviorally crippled. Finley sent the General to the Bronx Zoo when he reached adulthood. For a time he was said to be an extremely popular attraction. Then one day the General choked to death on a rubber band that had found its way into the condor cage. Maybe somebody shot it at the bird, thinking it would be funny.

Finley reacted, at least in print, as if he'd lost a son, writing a florid eulogy to the bird whose "wrinkled pate and flabby jowls, with the toothless expression of a toothless old woman, led the imagination back to some mysterious creature of the prehistoric past." Apparently, he didn't regret the decision to make this bird his pet and saw no link between the work of bird collectors, like him, and the desperate straits the wild condor now appeared to be in.

———

Sanford R. Wilbur, an archivist and biologist who worked with California condors in the 1970s, once tried to find the skins of every condor killed in the name of science. Wilbur says he started out by sifting through the records kept by first-rate collectors such as the late Joseph Grinnell. References he found in those records led him to a lot of libraries, where he sifted through more records and read lots of microfiche. References to kills turned up in hundred-year-old journals and obscure scientific publications. After weeding out some of the tallest tales, he started looking for cross-references. When he was done, he had a documented record of three hundred condors known to have died between 1782 and 1976, and a list of forty different condor eggs taken from their nests by collectors. Eighteen of the deaths were the result of natural causes such as disease. Another thirty-five had no known cause at all, which means collectors might have been involved. The remaining 247 condor deaths were the direct result of human intervention, including forty-one malicious shootings, 177 shootings by museum collectors, three deliberate poisonings, and twenty-six other human-caused deaths.

Finally Wilbur went to see most of these condor eggs and skins. He ended up in the basements and back rooms of homes and museums, pulling open drawers that didn't look like they'd been opened for decades. "Many of the skins I saw were pretty ratty-looking," he said. Specimens of condors on display at the museums were sometimes painted to obscure their age.

Wilbur says he also found a total of seventy hollowed-out condor eggs, including one that turned up in the trash behind a museum. Wilbur thinks the skin collectors helped bring the condor down by doing most of their collecting all at once, at the turn of the last century. The skin collectors themselves would have bristled at this claim, he says. They would have blamed the people who took the condors' eggs.

seven

EGGMEN

═══

In the various museums and private collections of the world, there are perhaps not more than fifty [California condor] eggs in existence, while there are about seventy eggs of the Great auk, a bird now extinct. It is not surprising then that [egg] collectors have been offering big prices.

—W. L. Finley, "The California Condor,"
Nature, August 1926

I have a picture of an egg collector on my desk—a black-and-white shot of a grizzled old farmhand known as Kelly Truesdale. He's sitting in the corner of a dim shack in an unknown location, wearing a dirty work shirt and pair of dirty jeans. His stone-faced expression reminds me of Buster Keaton. He knows what the joke is, but he'll never admit it.[1]

A flat wooden tray full of birds' eggs rests in Kelly Truesdale's lap. Trays just like it are stacked up in a cabinet behind him. I don't see the condor eggs Truesdale was so good at finding, but that's not surprising: Truesdale would have sold them to the highest bidder before they were even laid.[2]

This picture was taken in the 1950s, when Truesdale was a bent old codger and egg collecting was a badly tarnished occupation. But it was not always thus. From the end of the Civil War to the start of World War II, egg collecting was a mass obsession in the United States, a hobby in the sense that heroin is an analgesic. In *A World of Watchers: An Informal History of the American Passion for*

Birds, the historian Joseph Kastner called this craze a side effect of the rise of the leisure class and of the American magazine, but collectors always said the egg came first. They wrote of farm boys "seduced from the furrow" by Mother Nature's "painted oval souvenirs," and of grown collectors with a language all their own:

> The shapes are defined as generally elliptical, long elliptical, short elliptical, sub elliptical, long and short elliptical, spherical, oval, short oval, long oval, pyroform (one end pointed, the other broad), long pyroform and short pyroform. . . . The patterns on the shell surfaces are described as wreathed, capped, overlaid, scribbled, scrawled, speckled, streaked, marbled, spotted, dotted, splotched, splashed. The infinite gradations of color [include] tawny olive, greenish glaucous, aniline lilac, Quaker drab . . . odd facts the other birders miss.

The giant pale bluish-green egg of the condor was the ne plus ultra of the collector's world. Men whose collections numbered in the tens of thousands dreamed of laying their hands on one, and most were ready to pay top dollar. Egg collectors living in the condor's rangelands learned to cover their tracks and keep their mouths shut. It was always wise to act as if competitors were lining up to stab you in the back.

Those things happened from time to time: condor-egg collecting was a discipline that had a shady side. A hometown friend of Truesdale's, who understood why he chased the condors' eggs, once wrote about how some of Truesdale's competitors appeared to swim in the "intrigue that seems to characterize oological activities." In his book *Man and the California Condor,* a local rancher named Ian McMillan said Truesdale learned that the hard way, when he agreed to take a friend to a cave that had an egg inside it.

When the men arrived, they decided to take some pictures of the scene, but they hadn't thought to bring a camera. Truesdale knew there was a camera store in the town of Paso Robles, which he could reach in two days. Leaving his friend to guard the camp and the egg, Truesdale hiked down out of the mountains and caught a passing stagecoach. Five days later, back at the camp, Truesdale discovered that his friend had packed his things and left, taking with him the cotton-filled coffee can in which Truesdale stored his eggs. Climbing to the nest, he saw scattered wisps of cotton and not much else: the egg he'd he planned to steal from the birds had been stolen by his (former) friend. "With no one in sight he shouted," wrote McMillan. "But there was only silence from the surrounding chaparral."

Truesdale met his share of rotten luck, said McMillan, who wrote that the collector once flushed an adult condor from a cave before seeing that the egg was sitting on the bird's stubby feet. When the condor jumped, the egg flew forward past the man, exiting the cave and rolling off the edge of a cliff.

Scenes such as those were very rare and very hard to verify, according to the archivists and ornithologists who dig around in egg collectors' records. One of the reasons is that it's all but impossible to find a condor's cave, let alone one with an egg inside. Once, while I was in Arizona, a biologist pointed toward what she said was an active condor nest—a cave in the middle of a wall of red rock. After staring at the spot for five or ten minutes, I decided she was joking. But just as I was giving up, an adult condor slid out of a crack onto a narrow ledge, like paper coming out of a printer. When it slid back in, I still had trouble seeing where the bird had come from.

Even if you were to spy a pair of condors sitting on a ledge with a sign that said THE EGG IS OVER HERE, getting to the egg was a job

best left to locals with a death wish. These caves were impossible to
reach, and the birds were good at making sure they weren't followed. "No one ever found a condor egg on purpose," said Lloyd
Kiff of the Peregrine Fund. "You had to be good and then had to get
lucky."

Writers for egg-collecting magazines couldn't make much of a
struggle pitting humans against condor eggs. Many tried to com-
pensate by playing up the obstacles that stood between the hunters
and their prizes—"terrible precipices in whose sides the nest caves
of the great vultures were hidden."

Kelly Truesdale starred in some of the best of this pulp nonfiction. In 1911, he invited William H. Dawson, a prominent ornithologist and avid egg collector, to join him for a trip into the
coastal mountains in San Luis Obispo County. The goal was a cave
that had once given Truesdale an irreplaceable prize: a condor egg
that was a ghostly shade of white and not pale blue. Truesdale had
arranged to sell the egg to a collector in the East for what would
have been a record price, but the deal fell through when a middleman declared the egg a fake. Truesdale protested mightily, knowing
that the charge could wreck his reputation. When the middleman
refused to back down, Truesdale went to plan B.[3]

That was William Dawson, who agreed to follow Truesdale to
the same cave to obtain a second egg, which with any luck would be
the same shade of white.

Dawson, who wrote about the expedition in *The Birds of California* (South Mouton Co., 1923), promised not to bore his readers
with "the arduous details of that climb, and of our sufferings,
poked, prodded, buffeted and gouged, as we made our way upward
through the all but impenetrable thicket of buckthorn," and then
up a "half-cylinder shaped rock wall . . . stately and frowning not
only, but full of rifts and caves, soft places in the sandstone, scored

out by the elements, or once occupied by a softer substance now decayed and leaked out."

The condors saw the two men coming long before the men saw them. When Dawson noticed the great bird "soaring over the heights of his ancestral castle," the bird "is already looking down: soon [he] settles in at the top of a pine tree where we can study him with binoculars and telescope. We have a pretty good idea that his optical apparatus is better than ours at that, for he is ill at ease and presently casts off."

The men kept climbing, pausing when a second condor floated their way, and then a third. The birds circled slowly, then one of them turned and zoomed straight into a cliff. Truesdale and Dawson pointed the binoculars toward what they thought would be the mouth of a cave, but the cave didn't seem to be there. Both men had seen the bird clearly enough to know that it was up there somewhere, but all they saw was a jagged rock wall.

They kept looking. A long time passed. "Finally, Kelly caught a flash of color at the mouth of an obscure hole up the cliff-side. He called me over and I confirmed it—the head of a condor thrust anxiously forth from the mouth of the hole, and then withdrawn—a hole so small that I should not have looked for a falcon in it."

Dawson wrote that Truesdale flushed the condor from the hole by emitting a "current of catcalls," which sounds like a funny thing to do. Maybe he dared them to come out and fight like birds. Maybe he just yelled that he was back.

Truesdale was neither the most prolific condor collector nor the most eccentric. Lloyd Kiff of the Peregrine Fund says that honor goes to a climber named George Harris and his brother Jim, who may have taken thirteen condor eggs from three extremely isolated nest caves between 1889 and 1905. The Harrises worked for an egg dealer and collector named H. R. Taylor, the editor of a widely read bird-egg magazine called *The Nidiologist*. When the brothers brought him condor eggs, he'd let his readers know, in the hopes that one of them would offer him a huge amount of money. Kiff says $500 was the highest price ever paid for a condor egg. Rumors of higher fees have never been confirmed, and Kiff doesn't think they were received.

The Harrises did some stupid things to reach the caves that sometimes had those condor eggs inside them. Often they did little more than tie a single rope around a boulder at the top of a cliff. Then, without attaching safety ropes or harnesses, they'd grab the rope and scramble several hundred feet straight down.

Condor-egg collectors were eccentric people, Kiff explained. One collector lived in a cave he'd dug in the side of a cliff; another wrote that he had almost reached an egg when the cave he was in collapsed, sparing the man but crushing the prize. Yet another collector reached the entrance to a cave by tying one end of a rope around his waist and looping the middle around the base of a manzanita bush: the far end of the rope had been tied to a rock down at the bottom of a cliff that appeared to weigh as much as the collector. When this man stepped off the cliff, the rock rose slowly as he floated down to the mouth of the cave. When he had the egg, he stepped outside the cave, into the air—and more or less floated back up.

Truesdale did turn out to be the best-known member of this group. Ian McMillan and his brother, Eben, saw to that. Both rode shotgun on a few of Truesdale's early egg trips, soaking up everything the older man cared to tell them about the condor and its range. "We were studying the principles and the workings of ecology," Ian McMillan wrote in his *Man and the California Condor*. "[This was] long before the new science was heard of in the condor country."

I never met Ian McMillan—"Ike" McMillan to his friends—but I did meet his brother, Eben, one fine spring day in the 1990s. I was working on a condor story for National Public Radio. I'd just read about Kelly Truesdale in big brother Ian's book.

"Why don't you come over for some lunch," Eben said when he heard I was in the area. "Take the road you're on right now until you see the tree with the big metal sculpture next to it—that's the tree James Dean ran into on the night he died, you know. Anyway, when you pass the tree, turn right and just keep driving. The ranch house is down the road."

We talked for at least two hours on his shady front porch. Actually, *he* talked, but that was fine with me. Eben McMillan was a folk hero at that point in his life—reporters and environmentalists dropped in all the time—but when I was there he didn't want to talk much about himself. What he did instead was talk a lot about Truesdale.

"Truesdale and Dawson didn't make it to the cave in time to get another one of those white eggs," he said. "The chick had already hatched. But Kelly took the eggshell pieces back with him, and they were just as white as the first egg. Kelly ended up selling that first egg for something like two hundred and fifty dollars, which was the highest price ever paid for a condor egg. You could buy a nice used Model T with that kind of money. Kelly wasn't near as dumb as he looked," McMillan added at the end. "He knew the land just about as well as the condors did. He understood how it fit together."

The egg-collecting boom went bust after World War I: young collectors no longer rushed the nests en masse in the spring, and the experienced ones started getting old. Rich enthusiasts hunted for condor eggs at estate sales; some purchased whole collections. Laws that made collecting rare eggs a crime had begun to proliferate, and the scientific skin collectors had abandoned their old allies.

Condor-egg collectors were among the lowest of the low in the minds of sentimental bird lovers. Egg thieves who did it for the money would have been the lowest of them all. "The taking of eggs for the purpose of selling them for a few paltry dollars [is] an outrage," wrote photographer William Finley. In his view, the most outrageous of the egg thefts were "perpetrated under the guise of collecting for scientific purposes."

Many of the condor eggs were either lost after that or treated with a weird respect. One collector is said to have carried his egg in parades. One missing condor egg was found inside an otherwise empty Quaker Oats box. Another was hidden for years under a sink. One ended up in the attic of a Texas tire store; its owners said it was worth at least $10,000.

Ian McMillan didn't think the aging egg collectors deserved the grief they were getting. It was the men with guns who were the real problem, he insisted. "Prowess with a gun was a mark of high distinction," wrote McMillan. "In this philosophy there was no concern for destruction or depletion. . . . The more uncommon and rare the target, the more quick and eager was the shooting."

———

Condors looking down on California in the 1920s may have wondered what was going on. Big metal birds that made an endless roar flew back and forth across the mountains. Long metal animals that belched black smoke rolled back and forth on gleaming trails. Former feeding grounds had been replaced by endless lines of fruit trees.

The first recorded fight between a condor and a car took place in the 1920s. The bird was devouring a carcass on the Tejon Ranch when a rancher decided to chase it in his jet-black Model T. As the car approached, the condor tried to turn and fly away, but it had just eaten and couldn't get off the ground. The Model T followed when the bird scrambled off, futilely running and hopping and flapping its wings in an attempt to build speed. When the car got closer, the bird threw up the meal it had just eaten. When the condor slowed down, the car swerved past it, turned around, and headed straight for the bird. Just as it was about to hit the condor, the Model T stopped and a door flew open. The man who got out happened to be, in addition to the ranch foreman, a former ostrich wrestler. When the condor tried to run away again, the man lassoed it and tied it to the back of the car:

> And as he drove back to headquarters thirteen miles away his prisoner kept a violent metallic protest by wildly swinging the tire chains he found in the back of the car. The loudly heralded arrival of [the foreman] and the condor collected most of the ranch personnel, including the awed, superstitious ranch Indians. In the morning the condor was gone. The large wire cage in which the bird had been placed had been cleanly slit. After this incident the Indians took sly pleasure in reporting each time [the condors in the area] were seen soaring over ranch headquarters. The Indians insisted in reporting these flights that the condors dipped low in salute. . . . As they flew over, raising one yellow talon to each yellow beak in the familiar gesture of derision known the world over.[4]

The condor's range was shrinking fast, especially in the south, where the city of Los Angeles was spreading out like a pancake on a griddle. In 1870, Los Angeles was a dying cattle town with a total of

5,728 residents; thirty years later, it was a booming city of 102,000. Businessmen dreamed of Malthusian growth on lands to the north of the city, and on lands on the far side of the Transverse mountain ranges. Growth like that had long been kept in check by a chronic shortage of water, but in the twenties, everybody thought that problem had been solved. City engineer William Mulholland solved it back in 1915 when he formally opened the aqueduct that drained the Owens River Valley, sending water south across the mountains into Los Angeles.

If that boom had not gone bust, the California condor would have been doomed. That's just my opinion, but it seems like common sense if you think about the trend lines. The city of Los Angeles was shooting north toward the heart of the condor's domain. With it came more hunters, more hikers, and more rich folks who would try to buy the land. Houses would have popped up near the last of many dams built to hold the water taken from Owens Valley. This was the St. Francis Dam, which was completed in May 1928. It blocked a valley that drained into the Santa Clara River, which wound west toward the ocean past towns such as Piru, Fillmore, and Santa Paula. People would have bought into these towns and built them up before moving farther up into the mountains. Condors would not have been protected then. They would have quickly disappeared.

Everyone who's ever lived in these small towns knows why that future never happened: at 11:57 P.M. on March 28, 1928, the St. Francis Dam collapsed; forty-five million gallons of water began rolling toward the ocean, led by what at one point was a 180-foot tall wall of black water.[5] Houses, chunks of orchard, pieces of the dam, and bodies were also carried slowly forward by the flood. People who had cars were able to outdrive it. Those who tried to run were not so lucky.

Everyone I knew when I lived in Piru knew someone who was there that night. At the local grocery store these people all told stories that kept kids like me awake on moonlit nights like that one. I would sit in bed and watch the hands on the clock hit 11:57. Then I'd listen for the eerie rumbling sound, described by some of the survivors as the distant roar of an approaching train; the wind in a giant pair of wings.

eight

PATRON SAINT

===

The best thing that can happen to a condor nest is that nobody finds it.
—Carl Koford[1]

Somewhere in the Bering Sea, 1943: The military catapult roars across the deck of the USS *Richmond* like a giant sprung trap, whipping the top-wing scout plane toward the windward side of the cruiser. The part of the catapult that holds the plane accelerates to sixty miles an hour in the space of fifty feet and then hits a padded brace and stops dead. But the single-engine Kingfisher equipped with three machine guns and one depth charge keeps on going, sailing off the deck and out over the water.

The jolting launch makes the pilot and crew of the plane feel like human cannonballs, but they're used to it. They know that when the plane leaves the catapult behind, it will start to lose momentum, but only until the propellers seize the wind. At that point the scout plane will come to life and arc up into the sky. As the plane rises, Carl Koford will take out his maps and his binoculars and begin looking for Japanese warships. As he did this surveillance work, he must have thought about the bird he once described as the "acme of soaring flight." After all, this was the man who would later produce one of the most influential and controversial endangered species reports ever written: *The California Condor*, researched before and after World War II and published in 1952.[2]

The outstanding characteristic of the flight of condors is high stability in soaring. Frequently even an experienced observer mistakes a distant transport plane for a condor or a condor for a plane.

Carl Koford is the patron saint of condor field research. During World War II, he was the barrel-chested kid in the rear cockpit of the navy scout plane, in between the pilot and the tail gunner. It was his job to scan the water for the periscopes of submarines while looking for potential bombing sites along the Alaskan coast. When he saw a vessel on the Bering Sea, he pointed it out to the pilot; their plane would circle down to buzz the craft until it raised the proper flag. Koford also watched the skies above the horizons for thin black dashes that could only be approaching planes: if the dashes you missed were Japanese Zeros, you could be blown out of the sky. Top-wing scout planes were easy targets compared to Koford's birds.[3]

The pursued bird dives downward in a steep flex-glide, twisting from side to side, and the pursuer follows. Both swoop up out of the dive, flapping at about the same time, flapping at the end of the swoop to gain every possible inch of altitude.

No one knows what went through Koford's mind when he was flying those surveillance missions. He was not allowed to even mention his work in letters to family and friends, and when the war was over he didn't seem to see the point in sitting around telling old war stories. But when the world's leading expert on California condors looked down out of his plane for signs of trouble, it's a good bet he thought about the vulture. Koford had been living in the company of condors when Pearl Harbor was bombed. When the war was over, he picked up where he'd left off, taking incredibly detailed notes on the lives of condors in the wild.

In pursuing the field research I observed living condors between March 1939, and June 1941, on approximately 400 days. After a period of service with the United States Navy I watched condors on 80 days between February and July 1946 and on 15 days subsequently. The record of my personal observations consists of 3500 pages of field notes.

Carl Koford's study gave biologists a voice in fights affecting the future of the species, partly by providing them with a legally defensible set of scientific observations. After Koford, it wasn't enough for ranchers to insist that condors sometimes flew off with living cattle, or for hunters to insist that the birds would soon be gone regardless of what they did, or for the developers and agribusinessmen who wanted the condor's land to simply take it. After Koford, they needed proof. When they failed to offer it, the condor's defenders had a way to make them retreat—they had the "definitive" study of the bird, and they weren't afraid to use it.

Koford's book changed the political course of the condor wars by insisting that the bird was savable. Even more important, the study redefined the condor as a creature whose survival was tied to the fate of the wild. Koford's condor was a bird that needed isolation like fish need water. These birds could not be protected by rules that limited logging and hunting and fishing at the heart of their range; they needed refuges that were closed forever to humans, period. Koford wasn't the first to put this argument forward, but he was the first to try to back it up with facts he found in the field.

The need for these facts rose sharply at the end of the 1930s, when the U.S. Forest Service tried to build a road through a stretch of central California wilderness where condors bathed near waterfalls and nested in caves. When the road-building project was delayed by a proposal to turn the area into a sanctuary, the people who wanted the road called the bird an enemy of progress and a threat to the American way. "What Price Condor?" began an article that ran in *Field and Stream* in 1939. "The bird with the greatest wing-span has outlived its time."

The author of this article was H. H. Sheldon, an exasperated businessman and naturalist who thought the road should be built. Sheldon didn't think the condor deserved the attention and support it was then attracting, given the ominous state of world affairs. Europe was a "powder keg" with war "sparking along the fuse"; Asia and Spain were places where bombs were "blasting children to bits." A second world war was lurking out there somewhere, Sheldon feared, and of course he was right about that.

But when Sheldon read the papers in towns such as Santa Barbara, he didn't find articles warning of war. Instead he found headlines warning of a threat to a big black vulture that was doomed in any case. "In size and structure the condor is a magnificent bird," Sheldon wrote. "But its habits are deplorable and its purpose is finished." Sheldon later qualified the line about the condor's magnificent size, reminding the readers of *Field and Stream* that "size alone is no guarantee of virtue. If the elephant had the habits of a hyena, no one would mourn its passing."

Sheldon threw every insult at the condor he could think of, describing it as ugly, putrid, clumsy, obsolete—a downright "evil-appearing bird, dressed in a scrofulous black with bloody head":

> His habits would make a guillotine look like an angel of mercy. He is not a killer; he is a glutton of death. He displays all the characteristics of a pig, and some that the most disreputable pig would disown. Gourmand and ghoul, gorging himself on dead or dying animals is his sole object in life.

When he has stuffed himself to the limits of his capacity, not even his great 10-foot spread of wings can lift him from the ground without tremendous effort.

Sheldon did not understand how a bird like this could take precedence "over the siege of Madrid and [the saga of] Americans stranded in Shanghai." And "[f]ew in the East have heard of it. Few in the west have seen it . . . And so little is known about it, even in the west, where it has lived through ages, that conservationists actually believe it can be saved from extinction by setting up sanctuaries for its use."

Sheldon didn't think sanctuaries would do the bird any good, all but daring those who disagreed to prove their point. How sad it seemed to Sheldon that after failing to lift a finger to conserve the California grizzly, the state's "conservationists" would end up fighting for *this*. "I am a naturalist and a conservationist," Sheldon wrote, "and [I] believe the passing of any species to extinction would affect me with more regret than would assail the average individual. But to set aside a sanctuary in the belief that the condor will continue to exist is to act without knowledge of the facts." What Sheldon left out was that there were no facts about the wild condors' needs, because no one had ever sat and watched them live their lives.

Koford was living in the mountains that were the condors' home then—a place called the Sespe, because it had once been part of the vast Sespe Ranch. The research grant that put him there, funded by two rich condor enthusiasts, required him to work alone and at a distance from the condors, save for the occasional photo. His goal, as he put it, was to "discover, investigate and record *all* obtainable . . . data dealing with the natural history and especially the environmental relations of the California condor."

This project was the brainchild of Joseph Grinnell. This was the same Joe Grinnell who'd fought attempts to limit "shotgun ornithology," but by the 1930s, it was clear that these birds needed a different kind of attention. Grinnell didn't think there were more than twenty-five pairs of California condors left in the world in the late 1930s, and it was his hunch that those numbers were falling. Farmers and ranchers had been killing off ground squirrels and other rodents by spreading a slow-acting poison called thallium across a good part of the condor's range. Grinnell thought the practice was insidious but hadn't been able to stop it.

He asked the National Audubon Society to help him cover the costs of the project, and the society jumped at the chance. Audubon was then the most powerful environmental group in the United States, but in the early 1930s, it had been consumed by in-house policy fights and bitter power struggles. John Baker, a Wall Street investment broker who was Audubon's director, was trying to harness the organization's wasted energy when Grinnell got in touch with him. Baker said he would gladly add a condor job to a short list of projects to be covered by a brand-new Audubon Research Fellowship program. The ivory-billed woodpecker and the California condor would be the first two species studied.

Grinnell had one request. Audubon was "not to issue any publicity in relation to the California condor without submitting the same in advance for approval or rejection to those in charge of the research project at the University of California, and vice-versa." This was supposed to make it harder for unscrupulous collectors to find the nest caves. But the real goal was to drop a cloak of invisibility over the entire refuge. Grinnell apparently thought the condor could be saved on a need-to-know basis. Carl Koford's job was to find out whether he was right.

Koford hitched his first ride into condor country in the spring of

1939, wearing the hobnailed logging boots he always took to the field. In his backpack was a letter of introduction from Grinnell, who was known to everyone in California who cared about wildlife. The letter said Koford was a man with a sensitive and unusual ecological mission, for "it is the knowledge of the *living* condors that he specially seeks":

> In carrying out his field work Mr. Koford has been earnestly enjoined not to disturb the birds in any vital way; his aim is to practice technical "watching" with glasses, from a distance, whereby he will gradually learn the ways of life of this dramatically interesting bird species.
>
> I hereby bespeak for Mr. Koford the help of Forest officers throughout the country that he needs, in his work, to penetrate; also, the sympathetic and possibly outright aid of whomever else he may meet. . . . He desires no publicity whatsoever; none of us concerned wishes anything said or done, as through newspaper channels, which would in any degree increase the hazards of existence for these birds.[4]

The famous naturalist never said why he picked Koford to do this particular job; Grinnell died of a heart attack in 1939. "I trust that you will have notified one or two of your colleagues to watch out for him," said Grinnell in one of his last letters, to a friend in the U.S. Forest Service. "He is a quiet, earnest chap and will 'wear well,' I predict." Grinnell thought Koford might enjoy working in an isolated setting.

As it turned out, "enjoy" was not the word. Koford took to condor country like a feral cat with a notebook in its paw, stalking the birds for weeks on end, writing down everything. He always used a German technical pen with an extremely fine point. He always used one particular kind of notebook. He always copied his field notes into a second notebook before going to sleep, in script that's hard to read without a magnifying glass.

Field notation is a hoary art that greatly predates Charles Darwin, who started dividing living groups by species in the eighteenth century. But Koford wasn't looking for phylogenetic distinctions in the Sespe, or in finding a bug he could name after himself. What he did instead was to fill thousands of pages with descriptions of condor behavior. Hardly anybody studied so-called nonessential species in the 1930s, and when they did, they usually studied carcasses. But there was Koford, trying hard to write it all down. Wide-angle note taking of this sort was known as "the Grinnellian method," in which "the behavior of the animal is described and everything else which is thought by the collector to be of use in the study of the species is put on record at the time the observations are made in the field." If the day is overcast, you write that down. If the bird starts blinking, you start counting.

4:30 P.M.—This condor, like others I have watched, blinks constantly; most blinks are from a half to three seconds apart; 5 seconds seems about maximum. I wonder whether a red iris has any red filter effect on a bird's vision. The brightest orange on a condor is between the bill and the feathers between the eyes.

Koford was the first to note that parent condors rarely fly directly to the nest caves, choosing instead to land nearby and look around for predators. Instead of merely noting that a bird has landed, he writes about "a condor circling with legs dangling about 150 feet above the cliff," and then touching down after making five quick backward movements with its wings. After this bird landed and opened its bill, Koford noted an "orange tongue lying on the lower mandible"; a few seconds later "its head gave one sharp shake as if to dislodge a fly."

"I have never seen his equal," says Steve Herman, a staunch defender and former student of Koford's. "Biologists around today say a lot of it is hype, but let me tell you something. In the field, they wouldn't have raised a pimple on Koford's ass."

━━━

November 2001: I'm sitting at the point of a rock escarpment in the middle of the Sespe Condor Sanctuary in the Ventura County backcountry, taking in the prehistoric view. The map in my pocket says I'm looking at one of the lines of mountains in the Transverse Ranges. But I'm having trouble seeing the "line" part. What I see instead is a jumbled mess of hugely varied landscapes, bent and broken in a way that makes it look like something punched its way up through the crust of the earth.

Off in the distance, near the horizon, is the brown haze that marks the outskirts of the greater Los Angeles metropolitan area. Closer in, on a grass plateau cracked open by earthquakes, I see tilted pasturelands once grazed by Spanish cattle. Rolling hillsides end in cliffs that make it look like a colossal beast beneath the earth is trying to punch its way out. "Geologic upthrust" is the scientific term used to describe the scene in front of me. I half expect to see a giant stone fist come crashing up through one of the mountains.

I can see a knifelike fissure running roughly north and south. I'm told it's an offshoot of the "big bend" that turns the San Andreas fault line to the east beneath the Transverse. When David Brower joked that condors could only be saved by an earthquake big enough to put Los Angeles under the Pacific Ocean, it was this network of fault lines that he was counting on to do it.

But the fissure isn't what dominates the view from the spot known by biologists as "Koford's Observation Point," or "Koford's O.P." The dominant thing is a pockmarked curtain of yellowish cliffs that ends in a wide plateau on my side of the fissure. The cliffs fall abruptly for several hundred feet into a valley full of chaparral. Koford's escarpment rises on the other side of that valley.

I'm thinking three thoughts while I'm taking in the view. One is that it's easy to see why Koford came up here all the time. The pockmarks in those yellow cliffs are nest caves used by condors in his day. From here, with a sitting scope, he could sit and watch them all day long.

I'm also thinking about the grizzly bear head that used to hang on the wall of Lechler's grocery store in Piru, at the southern base of these mountains. I remember sitting in Lechler's store in the early 1960s and staring at that head, imagining that its giant body was sticking out of the other side of the wall. When Mr. Lechler told me that was not the case, I decided the body was still walking around in the mountains to the north, feeling around for its head. That thought returned to me this morning when I hiked across some big black bear tracks.

I'm also thinking that you'd have to be nuts to try to move around out there. The chaparral below looks like concertina wire, and the cliffs resemble a scene from Mordor in *Lord of the Rings.* Moving back and forth would have been hell, which may be why Koford thrived there. Herman says Koford appeared to enjoy leaving students in his dust. And like all of Grinnell's disciples, he was skilled at living off the land. He drank the water he found in potholes and he often shot his meals; the speed with which he skinned small birds was on a par with the naturalists of old.

"Once when we were in the mountains of Mexico looking for the last of the Mexican grizzly bears," Herman wrote in an e-mail, he saw me struggling with a small sparrow I had shot, trying to relieve it of its skin and stuff it in a way that would preserve it. . . .

Carl watched me . . . for a few minutes and then took the bird. He fit his drugstore reading glasses on his nose and settled into a canvas chair and began wielding his scalpel with considerable skill. Zip, zip, zip and the skin was off. A few more minutes and the bird was stuffed and wrapped, as if it were lying in state. Something that would have taken me nearly an hour had taken Koford minutes.

Old-school ornithologists like to joke that you can't really understand a study species until you've eaten it, but condor steaks were not on Koford's menu, and the smaller birds were eaten rarely. Koford's old friends say he often skipped meals when he was out in the field. At times his diet seemed to consist entirely of canned apricots. One of the biologists who followed Koford into the Sespe Condor Sanctuary says he could tell where Koford had been by looking for the empty cans.

Koford had a cabin a couple of miles to the south of his escarpment. It looked out on a pond the condors used to bathe in all the time. But the bulk of his work was done on the other side of the broad dry canyon interrupted by the curtain of pockmarked cliffs. After hacking partway through that nasty chaparral for a couple days, he'd build a blind and watch the condor caves for weeks at a time, focusing on a breeding pair with a fledgling he called Oscar. Koford saw Oscar's parents chase off ravens and golden eagles. Once he watched the chick try fruitlessly to scratch its itchy head. "Five times the left foot was brought to the head to scratch," he wrote in his field notes. "Each time only one or two quick strokes were managed before the foot had to be hastily replaced . . ." Koford also watched a parent bird feed the chick by holding its open bill about an inch above the chick's head. "The chick then jams its head up into the adult's throat from below and the adult's head [starts shaking], either from regurgitation or from the wrestling actions of the chick. After a few seconds the chick pulled its head down out of the parent's throat, holding a light-colored chunk of partially digested animal remains. The chick wolfed it down and beat its wings to beg the parent bird for more. It went on this way for quite some time."

Koford didn't think much of the gizmos sometimes used to make the lives of the note-takers easier. Other field biologists tape-recorded their observations and transcribed them later. Koford thought the practice lazy. His 32-power sighting scope and the wreck he called his car may have been his only prized technological possessions.

"Carl was frugal," said Herman, putting it delicately. "For instance, his car had a tendency to stall. . . . It turned out that he had adjusted the carburetor so that the gas/air mixture was very, very lean, i.e., as little gas as possible relative to the air. It was so lean that the motor only ran when the car was moving."[5]

Herman, who is in his seventies, thinks the differences between what Koford did and what field biologists do now are extremely difficult to fathom. Koford never tried to trap the birds and bolt ID tags to their wings or test their blood for man-made poisons. He never tried to follow them with tracking devices or even to find out where all the nests were. For the most part Koford sat and watched and wrote it all down, even if it didn't seem important.

"He was a generalist," said Herman. "He's the one who built the baseline. The fact of the matter is that condors really were wild birds in Koford's day, and even my day. They are no longer, and in fact they are about as far from being truly wild as anything could get and still fly around."

Herman wrote those words years later when the hands-off school of condor management launched by Koford's work was under attack by so-called hands-on scientists with the zoos and the federal government. He thinks abandoning the Koford approach was a terrible mistake.

━━━━

Koford wasn't always the only human near the condor caves. Before the war, he chased off strangers armed with cameras and guns, but sometimes the strangers made it past him. Once, a magazine photographer brought a model into one of the nest caves, shooting pictures while she posed with one of the birds. Others would flush out the birds by throwing rocks at them or firing shots into the air.

Some who came to visit Koford and the birds were anything but stangers. Loye Miller made the trip in 1939 with his son Alden, who'd just taken over the job of running the Museum of Vertebrate Zoology at U.C.-Berkeley. After peering into Oscar's cave through Koford's sighting scope, they watched a pair of condors stage what Loye Miller called a "clumsy dogfight" aloft. He also wrote of watching a group of eleven birds fly near the ridge the men were standing on, passing back and forth at eye level and striking "a variety of poses in the air." Miller was amazed by the way the birds moved around in air that was almost completely still, twisting their tails and calibrating the giant black feathers at the ends of their wings. Later they saw the condors linger over a scoured carcass, appearing to "loaf on the wing for a time for the mere pleasure of the exercise."

Koford spent a lot of time with J. R. Pemberton and Ed Harrison, oilmen and former egg collectors who helped fund his research grant and told him how to get to Oscar's cave. Pemberton was a physically imposing man who'd made a fortune building a railroad in Patagonia; now he was the California state official who evaluated oil-drilling proposals. Harrison was the heir to an oil fortune who didn't need to work for a living; when condor-egg hunting was outlawed in the early 1900s, he'd begun pursuing other egg collections. The two men shared an interest in the condor and the cracked terrain that was its home. There was lots of oil hidden under this terrain, and they were looking for potential drilling sites. When movie cameras first became portable, Harrison and Pemberton took one up the mountains to the high plateaus that form parts of the Sespe, where they started filming condors in their caves.

These two men were Koford's best friends in the field at the time, always bringing him groceries and the latest camera gear. On more than one occasion, the three of them filmed condors flying in and out of their nests, even though that wasn't always what the condors wanted to be doing. Pemberton sometimes encouraged the birds out by firing his pistol. Harrison and Koford filmed each other sitting inside nest caves with a condor in their arms. I saw the films in Harrison's office in Los Angeles.[6]

Koford quit his fellowship to go to war when the Japanese bombed Pearl Harbor. During the war, he may have used an old birder's trick to tell Harrison where he was. On postcards that were otherwise free of any hints of his location, Koford is said to have mentioned a species of bird that could only be found near the Bering Sea, where the USS *Richmond* was stationed.

━━━━━

World War II changed California almost as much as it changed Japan. Hundreds of thousands of military families moved out to the Los Angeles area; rows of homes and freeways were unrolled in every direction. Factories and cars put much more pollution up into the air. Farmers started spraying everything they grew with an

amazingly effective bug killer known as DDT, which was manufac-
tured by the ton in a factory near the Santa Monica Bay. Ranchers,
not to be outdone, started buying a new squirrel-killer called Compound 1080.

Koford returned to his study site in the mountains north of Los
Angeles to find that an oil-drilling platform had been built on top
of the pond the condors had once bathed in near his cabin. Access
roads to other wells had been built and then widened. Mining companies were lusting after phosphate reserves, and hunters armed
with army surplus rifles buzzed around like horseflies.

Koford noted the changes and went back to his birds. They
seemed less common than they'd been before the war—smaller
groups and fewer sightings—but there didn't seem to be a way to test
that observation. After interviewing ranchers and other locals and
watching the birds, he guessed that at the end of World War II, between sixty and 120 condors were left. Later, his allies only mentioned the low end of that estimate, which was very low indeed. It
meant that it would only take a small streak of bad luck to send the
species into yet another downward spiral, and with all the changes
going on in California, bad luck was certain. In other words, Koford
wrote, people should be told that "the precarious natural balance of
the population can be easily upset in the direction leading to extinction."

Koford thought his field notes proved the need to separate the
men from the birds. So did Alden Miller, Koford's adviser after the
death of Joseph Grinnell. Miller was an accomplished ornithologist
whose conservation strategies were rarely questioned and often accepted as gospel. In public, he was formal to the point of seeming
imperious at times; behind the scenes, he was known as an enemy
you did not want to make. Miller fought ferociously for the things
he held dear and true.

He didn't think there was any doubt about what the condor needed. Like Koford and Grinnell, he thought the condor needed untouched wilderness and absolute isolation. Activities that might draw attention to the bird were not to be condoned. Miller felt strongly that photography was one of those activities, arguing at times that too many pictures of the birds had already been taken.

This new tactic put Koford at odds with his old friends and patrons Pemberton and Harrison. But when the war ended, Koford met Ian and Eben McMillan, who would eventually do more to promote the hands-off approach to condors than almost anyone else. Ian McMillan wrote that he was doing chores in his yard one day when Koford drove up in his broken-down jalopy and introduced himself. McMillan saw a "lean but well-built youth who turned out to be older than he looked." Maybe it was the fine spring day or something in the way Koford moved that made McMillan think of old times: "The situation was somehow reminiscent of former days, when Kelly Truesdale would arrive at the old homesteading McMillan canyon with intriguing reports of his collecting adventures."

The McMillans and Koford became fast friends. The writer Dave Darlington said Eben seemed to revel in Koford's many eccentricities, including his habit of bursting into unexpected laughter and not telling anyone why. "He had a very peculiar sense of humor," Eben told Darlington, adding he never once saw Koford in a coat, even when the ground was thick with snow.

He was always practicing self-discipline; it would have been pretty near impossible to sell him something he didn't need. Koford was a totally objective person—he never joked, never said anything he didn't mean, never went out on a limb . . . he was sort of a mechanical person—he had no human weaknesses. His integrity was untouchable, which made him unique in a society where integrity has lost its meaning.

Eben and Ian McMillan helped Koford meet Truesdale and many of Truesdale's old competitors. Then they showed him how to get around. In the 1940s, no one knew the condor's range like the McMillan boys did, and they passed all that knowledge on to Koford. They fed him and took care of him, and defended his work when Pemberton and Harrison faded from his life, and when Koford left the country in the 1950s, the McMillans spoke for him.

People who have made their way through all 3,500 pages of Carl Koford's field notes say that his image of the condor seemed to change when he returned from the Pacific Theater. References to condors disturbed by people seemed to get a lot more common, as did the idea that condors were less sanguine than they looked:

One factor leading to a false idea of tameness of condors is the lag in reaction of the birds to disturbance. Commonly when a condor does not leave its perch as a consequence of a man close by, it will leave several minutes later when a man has walked several hundred feet away . . .

One man, by disturbing the birds at critical places during the day, can prevent roosting over an area of several square miles.

Koford said the condors were rattled by the sounds of trees being cut and the sounds of airplanes flying over their nest caves. "Even the buzz of a motion picture camera 100 yards from a perched adult appears to be noticed . . . ," he wrote. Still photographers were even more of a problem.

The failure of some nests known to me was probably due, at least in part, to the activity of these men. Even with great care, a party which I assisted kept the nesting adult from the egg or chick on some occasions. Other photographers were much less solicitous of the welfare of the birds and some of their activities were literally cruel. Even when photographing a bird with a large telephoto lens, one must be comparatively close to the subject. . . . There is little to be gained by attempting to obtain more photographs of these birds.

Koford's study was hailed as definitive long before it was published by the National Audubon Society in 1952. In 1947, the federal government turned his study area into a 53,000-acre condor sanctuary, tearing out the hiking trails and locking out the fishermen and hunters. Koford was delighted until he heard that the place would be called the Sespe Condor Sanctuary, which was sure to draw the birders in.

It was on the maps and that's where everybody thought they could go to see condors—including people like Roger Tory Peterson. He wrote in one of his books, "Well the condors have their sanctuary now." As though a big sanctuary would take care of everything, even though the condors spent most of their time off the sanctuary, foraging on private lands.

Koford didn't think *anybody* should be let near the condors, including researchers like him. Watching from a distance on a limited basis was probably okay, but getting close enough to be noticed was essentially an act of violence. Koford had also become convinced that there was no justification for entering the nest cave with a con-

dor still inside it. He had done it so many times, but those were
mistakes he did not want to see repeated.

This was not a notion that went over very well in the scientific
community; many critics say it helped accelerate the birds' decline.
Noel Snyder, the leader of the California Condor Recovery Program
for part of the eighties, called this philosophy "a curse that endured
for nearly 30 years, totally inhibiting much needed research."[7]

There's no doubt that Koford helped to complicate the lives of
the scientists who later tried to lay their hands on the last of the
free-flying California condors. On the other hand, the condors
would not have been out there if not for him. They would have
been gone.

nine

HANDS-ON

The wild wastes of a century ago are now dotted with lumber mills, mine shafts, and smelters. Under the earth extends a network of pipelines for oil and natural gas and above it, a network of high extension wires for electric current. The canneries and packing houses, oil refineries, aircraft factories, and movie studios ship their products to every corner of the Nation and beyond. The Californian of today feels a personal pride in the state's gargantuan public works: highways, bridges, dams, and aqueducts. And most of all, of course, he exults the region's "happy future."

—*California: A Guide to the Golden State*

O ne hot morning in the summer of 1966, my mom and dad told my sister, Kirsten, my brother, Peter, and me to get out of bed, put on our Sunday-school clothes, and walk down to the Piru train station. I knew right away that they'd gone nuts, since it wasn't Sunday and the train station had been closed for years.

Half an hour later we were riding east in an old-fashioned wooden passenger car attached to a gleaming steam engine that had been rented by the company my father worked for. When the train started rolling out of town, my Mexican friends rode their bikes alongside us, waving and yelling things that made me glad Mom and Dad did not speak Spanish.

This would be the last train to leave Piru for at least fifty years. When it was gone, the tracks were torn up and trucked off to a dump. But I didn't know that at the time, and the ride itself was fun, with booze for the grown-ups, food for the kids, and entertainment out the windows. Mom says there was a jazz quartet in one of the passenger cars, but that must have been a different ghost train. I don't remember hearing any jazz.

Halfway to our destination, a bunch of fake Indians rode out of the orchards on some horses. Fake cowboys were chasing them, shooting fake bullets.

The Indians escaped by riding up the road that led to the county dump. We chugged eastward toward the former hog farm that was our destination, passing a notoriously dangerous motorcycle track and a work crew from the county jail.

The train ride through the valley was supposed to celebrate the end of an era, I was told—a time when the residents of Los Angeles proper called the towns at the edge of the Transverse Ranges "the home of pigs and prisoners." As the train trip ended, we looked out at the beginnings of the master-planned community my father had been working on. In real-estate speak, the place was far too big to qualify as a suburb, being better suited to phrases like "semi-autonomous regional center," "new town," or "middle landscape."

The city was called Valencia. I remember noticing that Dad was proud of what he'd helped accomplish. I also remember meeting the mayor of the *real* city of Valencia, who'd been flown in from Spain to mark the occasion. Later that day Peter was run over by a golf cart; when he turned out to be fine, we caught a ride home to Piru.

The historian Carey McWilliams once wrote that the history of Southern California is the history of its booms, and in the mid-to-early sixties, Los Angeles blew up like a bomb. The freeway to be known as Interstate 5 was under construction then, and suburban developments of every shape and size were chasing it north. These were changes you could see from parts of the condors' breeding grounds.

Ian McMillan, the rancher and well-known condor activist, thought of suburbs as abominations. In his book, he wrote about the fight to save the condor's world from the forces of suburban sprawl; he wrote of looking down upon their work one morning in May 1965 from the top of a mountain in the middle of the Sespe Condor Sanctuary. Pointing his telescope toward the homes, McMillan claimed to see what he always saw when he looked down from this spot: "rows of new dwellings that seemed to have spread across a few new acres of bottom land."

In 1965, the new town of Valencia would have looked as if it was rising all at once—homes, new shopping centers, an upscale golf course, and a Magic Mountain theme park. When Valencia was finished, this new part of California would be terraformed, clean and green, with no rotting cattle lying on the front lawns and no giant vultures in the trees.

McMillan didn't like the air pollution these new worlds produced. Up on that mountain in the spring of 1965, he'd been alarmed by the smog rolling toward him.

By late in the afternoon it was filling the canyons of the condor refuge to a height of three thousand feet. Looking in the direction of Fillmore I could barely make out landmarks five miles away. Similar landmarks ten miles away had been clearly visible this morning. As I looked out at this spreading blanket of foul air, I thought again of survival. What were the effects on the condor of this new man-made factor? To what extent would the change in visibility affect the bird's success in foraging? I had seen native trees that were dying from the effects of air pollution. I had experienced the eye irritation that attacks one in smog of high intensity. What is the effect of all of this on the telescopic eye of the condor?

Ian and his brother, Eben, spent part of the sixties trying to update Carl Koford's work. When they weren't doing that they were fighting a $90 million proposal to dam Sespe Creek, the largest undammed creek left in Southern California.

The creek in question happened to run through the center of the condor's range, crashing down out of the mountains through a twisted set of high rock walls before leveling out and slowing down just behind the town of Fillmore. "The crooked Sespe is fantastic beyond anything man has yet constructed," Ian McMillan wrote. "Even if it were not a main haunt of the California condor, the area would be worthy of care and preservation as an incomparable piece of wild country."

Strenuous endorsements of this point of view were provided by the New York office of the National Audubon Society, and by Alden Miller, the man who had succeeded Joe Grinnell as the director of the influential Museum of Vertebrate Zoology at the University of California at Berkeley. Miller left no doubt at all that he thought this plan would be a shot through the condor's heart. "There must be no yielding to development," he said, while pushing to expand the sanctuary. "We need to hold rigorously to what is now set out to be protected land."

The plan to tame the Sespe had been thought up with the help of the Federal Bureau of Reclamation, which was then in the process of losing its first major political battle to David Brower and the Sierra Club. Two large dams and one diversion dam were to block the flow of the Sespe Creek, filling two large reservoirs and one twenty-five-mile conduit. With water from the reservoirs, local developers would be able to build more houses. Towns such as Fillmore would bloom as hunters, fishermen, and water-skiers stopped in on their way up to the dams, and as the water stored behind the dams fed more farms and homes. The reservoirs would be stocked with fish pumped out of local hatcheries, the new road through the mountains would make year-round ingress easy, the fully developed project would draw two million customers a year, and Ventura County would live profitably ever after.

Local folks behind the Sespe project didn't like it when the National Audubon Society marched into town and denounced them as a bunch of condor killers. Fred Sibley's job had just been created by the U.S. Fish and Wildlife Service, and nobody seemed to know what he was going to make of it. What exactly was an "endangered species specialist" anyway? Was it like being the condor's public defender?

"Forty-nine dirty birds" is what the prodam forces called the condors early on. But when the fight to save Sespe Creek got hot, they abruptly changed their tune.

After briefly arguing that human needs must come before the needs of giant vultures, the backers of the Sespe plan began to insist that the condor would be lost without them. "Audubon is interested in saving nothing but the condor," wrote one supporter of the dams. "United Water Conservation group [named after the Ventura County United Water Conservation District] is a true conservation group. It wants to save vitally needed water *and* the condor."

Carl Koford's old friend Ed Harrison supported the plan to dam the Sespe Creek and the plan to build an access road through the condor sanctuary. Harrison insisted that the condors weren't nearly so reclusive and fragile as Koford said they were. He also thought the condor's repuation would improve if more "recreationists" saw them while driving through the "preservation zone," and surely this would make more people care about keeping the species alive. Approaching the birds could easily be prohibited, Harrison said. All you had to do was build tall wire-mesh fences on either side of the road.

Koford hated this point of view, as did everyone at Audubon.

But Harrison insisted that the old approach had simply failed to save the bird. He thought the time had come to take more drastic steps to save the condors from extinction: if that meant building fences and keeping birds in captive breeding programs, so be it.

As far as I know, there are no records of Koford's reaction to Harrison's plan. But it's safe to say he did not jump for joy. Ian McMillan was more outspoken, as usual, arguing that "with the advice and cooperation of the condor cagers, the developers [have] embroidered their proposals of becoming fairy godfathers to the condor."

The U.S. Fish and Wildlife Service threw biologist Fred Sibley into this fight in 1967 after Lyndon Johnson signed an early otherwise toothless version of the Endangered Species Act. With the money that signature freed, the service did something it had never done before: it hired a group of field biologists and sent them out to study the plight of four of the nation's rarest life-forms. One man was assigned to the black-footed ferret and another to the Everglades kite. A third was sent to Hawaii to report on several rare plants and animals.

The fourth assignment was condors. Hardly anyone applied. Eventually, it went to Sibley, who trained at Cornell University and was employed at the time by the Smithsonian Institution. When Sibley learned of the condor job, he was working on an island in the South Pacific, catching and banding albatross. When he heard that there'd been very few applicants, he called Fish and Wildlife to ask them why. Sibley says he was told that many had declined to apply because the job required expert climbing skills, which Sibley did not have. He applied anyway—and was quickly offered the job.

"I didn't hear the *real* reason there weren't many applicants until later," he said. "Turns out nobody wanted Alden Miller ruining their career."

Alden Miller died of a heart attack before Sibley got to California. When he died, Ian McMillan took his place on a scientific advisory committee formed to monitor Sibley's work. At the first meeting of this committee, Sibley laid out an extremely ambitious plan to assess and rank the threats to the condor's existence. First, he would find every current and former nest cave in the condor's range. Stray feathers, eggshell bits, and scrapings of feces would be collected at every opportunity, so that they could be tested for signs of malnutrition and poisoning from pesticides. Maps of distribution and abundance would be drawn up and updated. Attempts to count the condors would be staged on an annual basis. Birds would be organized. Koford's claim that the condor was allergic to humans would be put to the test.

Sibley promised the advisory committee that he'd stay out of active condor nests while the condors were inside them, adding that he wouldn't enter them at all when the birds were breeding, hatching eggs, or raising condor chicks.

All but one of the members of the scientific advisory committee thought Fred Sibley's plan was a sound one; Ian McMillan did not. "I was horrified," he said in an interview published in a book called *The Condor Question*. "I objected strenuously, and might have been a little harsh. But I felt I had to be. I was alone, absolutely alone, in the advisory committee."

Ian McMillan was removed from this committee at the end of its first meeting in 1967. For the next three years, he never stopped trying to bring Fred Sibley down, constantly questioning his findings, his research skills, and his honesty. When Sibley helped coordinate the first in a series of annual condor counts, McMillan said the numbers were wrong; when Sibley said the condors might not be finding enough food, McMillan called it a "fallacy."[1]

Outside condor country, Fred Sibley was known for both the strength of his fieldwork and the integrity of his findings. Jon

Borneman, a condor "warden" hired by the National Audubon Society, said Sibley was also known for speaking his mind. "He was one of those people who never had time for b.s.," Borneman said. "In that respect he was a lot like Ian."

Sibley and McMillan never got the chance to talk about becoming allies, but in many ways that's what they were. Behind the scenes, Sibley was a scathing critic of both the U.S. Forest Service and the Federal Bureau of Land Management, each of which controlled the lands that surrounded the condor sanctuary. "Even in [the best of times] the condor takes second place to development," he wrote in an angry letter to his boss back in Washington. "Unless the decisions are made at a higher level than district or forest, the condor is lost." In another letter to the Washington office, Sibley seemed especially critical of the way the local Forest Service was run. "Voluntary cooperation of the [two men in charge of] Los Padres National Forest appears to be impossible," he wrote. "Both worship projects and measure development in terms of dollars spent. Secrecy has long been a strong weapon and is still used with annoying frequency." Projects that might hurt the condor were approved and launched and then revealed to Sibley, whose angry responses were dismissed as moot because the condor was already doomed. This was an attitude the Forest Service tended to adjust to fit the exigencies of the moment. Roland Clement, a former vice president of the National Audubon Society, who was involved with condors at the same time as Sibley, told me that he once heard a local ranger insist that the condor could not possibly be endangered by a development proposal. When asked to elaborate, the ranger said he'd just seen a giant group of "baby condors" in a nearby canyon. Clement said it turned out that these "baby condors" were actually turkey vultures.[2]

Sibley understood why the McMillans held "the feds" in such

low regard. "They had ample reason to hate these people," he said. "They were never given honest answers. This was true to the end and prevented us from forming any sort of friendship."

Sibley did try to settle his differences with Ian McMillan once. "Jon Borneman [of Audubon] and I asked Ian if he wanted to have lunch," Sibley said. "He kindly invited us to lunch at his ranch."

When Sibley and Borneman arrived, McMillan showed the men around, talking about his work with California quail. Sibley was encouraged to inspect one of the roosting platforms McMillan built to protect the quail from night predation, which was a big problem then. Sibley remembers being impressed by both the platforms and by the graciousness of his host.

But the tone of the meeting changed dramatically when the steaks were served and the conversation turned to condors. Sibley said McMillan and his wife started yelling almost immediately, accusing their guests of sneaking into occupied nests and handling the condors. Sibley and Borneman denied the charge, but the McMillans did not believe them: "You're entering the nests! Admit it!" is what Sibley remembers. Not knowing what to say except, "No, we didn't," Sibley and Borneman ate and listened, thanking their hosts and saying good-bye when they were done.

"Ian's real shortcoming was that he could never discuss anything," Sibley said later. "He would just attack from second one—he didn't want to hear anything."

"It was like we were saving two completely different birds," said Borneman. "One was a fragile giant that threw up and flew away when it saw people; the other was a rugged, stubborn coot that would never *abandon* its young."

Sibley worked with that second condor. He said he had no time to worry about what the hands-off faction thought of his work. This was true in part because Sibley was almost always on the road, and in this case the word "road" is liberally defined. In the early 1960s, most of the roads that ran through condor country were narrow ruts at best: while hiking and climbing more than 850 miles in his first year on the job, he wore out three pairs of boots and broke two backpacks. "Fred was one of those people who does a day's work in about two hours," said Borneman. "He was just a persevering, dogged kind of guy."

Naturally, the condors seemed to watch his every move, according to Sibley's field notes. "In most instances the condor has soared low along a ridge and craned its neck to look at me as it went past," he wrote. "At other times the birds were seen circling once or twice [above me], or hanging motionless in the air to take a longer look."

Sibley didn't think he bothered the birds by hiking through their range. But the planes he saw while walking seemed to bother them a lot. Roughly a hundred planes and gliders flew across the Sespe on a typical day in the middle of the 1960s. Sibley said the condors would ignore an aircraft coming up from below, but would "react rather violently to one coming from above," twisting and turning like they do when attacked by golden eagles.

Once he watched a condor shoot forward out of a cave after hearing a sonic boom, but not to flee the area: skidding to a halt at the lip of a cliff, the bird started looking around for the source of the boom, watching planes fly overhead for hours. He saw other condors react in similar ways to gunshots, dynamite blasts, and passing trucks. Certainly, these birds were bothered by such intrusions, but Sibley wrote that he also saw a lot of resilience. These were birds that badly needed protection, Sibley wrote, but they weren't reclusive giants that would leave an egg behind to die.

These condors seemed much tougher than the condors Koford wrote about. That thought was enough to infuriate Koford and his allies, some of whom began describing Sibley as a tool of the prodam forces. This was false, to say the least, but Sibley didn't fight the charge in public—even in the wake of the lunch fiasco, he kept looking for a chance to reconcile quietly.

Those hopes vanished for good in the summer of 1967, when Sibley took an ailing condor to the Los Angeles Zoo for treatment. Ian McMillan immediately accused him of fabricating the bird's illness, and implied that the bird had been captured as part of a plot to circumvent the trapping ban. This was a claim that would never be dismissed, even though it was not true. Everyone who saw the bird in question knew it needed help, and at one point even Koford agreed that it needed to be taken to the zoo.

"It would have died otherwise," said Borneman, who helped capture the bird. "It was next to death when we caught it." Borneman added that the fledgling had been staggering around near a barn for several days when a fisherman called to report that it was refusing food and acting like it could no longer fly.

The next day, Sibley led a group of forest rangers and biologists up into the chaparral. *This won't take long*, he remembers thinking to himself. By the end of the day, the fledgling bird would either be dead from its ailments or recovering in the wild, full of medicine administered during a brief encounter with the veterinary experts at the zoo in Griffith Park. It would happen nice and easy, he thought.

But of course it did not. This was a condor that would soon be known for its habit of flying head-on into the chests of terrified zookeepers and visitors, sometimes knocking them onto their backs and standing on top of them; it also tried to breed with metal fences and tree stumps. To this day the bird in question holds the unofficial record for flesh chunks ripped from the arms, legs, and torsos of humans. It may have been the least approachable and in that sense "wildest" condor ever trapped.

Later they would name it Topa Topa and marvel at how tough it was. For two days, half-dead, the bird stayed just ahead of the scrambling trappers, often flying just enough to get to the center of a nasty stretch of chaparral, where it would collapse from exhaustion. When the men got close enough, the bird would rouse himself again and lurch forward into another nasty mass of thorny branches.

These dance steps were repeated for at least a day before Sibley noticed all the poison oak. He ignored the rashes and kept hacking through the brush, but others weren't so lucky: one biologist was rushed off to the closest hospital when his whole body seemed to start swelling. All of the trappers were raw and bleeding at the end of the second day, when the fledgling finally got too tired to move.

"We just rushed it and tackled it," Sibley said. "Stuffed it into a large duffel bag and took the struggling package down the road. One car stopped and asked what we had in the bag; Borneman yelled out 'ant lion' and the fellow drove off happy."

At that point it was too late to drive down to L.A., so Sibley took the condor to his house in Ojai and locked it in the garage. Before setting out on the trapping trip, he'd bought a huge mound of hamburger meat, just in case. But when the time came to feed the bird, he didn't know what to do. Should he scatter bits and pieces near the bird? Should he just drop the mound of meat and run?

Sibley said those questions resolved themselves as soon as he opened the cellophane wrapping on his emergency supply of hamburger. "The bird immediately struck at me," he recalled. "The neck pulled the head back and then whipped it forward, as a snake might strike. The bill hit the package, knocking it to the ground and before I could bend down to retrieve it [the condor] had consumed almost the entire package. In the next minute every scrap had been retrieved and consumed. What surprised me was the instantaneous recognition of the package as food. I'm sure the initial strike was hostile, but once the bill hit the food the whole scene reversed and it was a one-bird feeding frenzy."

The future Topa Topa was taken to the zoo, found to have nothing obviously wrong with it, and returned to the wild. To make sure it would eat something other than packaged meat, Sibley and some colleagues put a jess on one of the bird's legs and drove a stake at the other end of a long rope into the frozen ground. A deer carcass was placed nearby, so the condor could eat and still retreat.

Sibley began to relax, but that didn't last very long at all, because the condor wouldn't eat. Days went by while Sibley and his colleagues sat in the snow and watched Topa Topa sit and stare at the carcasses it was supposed to rip apart, failing to move when other birds moved in to eat. Sibley took a day off and went home at one point, only to receive a frantic phone call. The tethered bird had flown down to the carcass to start eating when two other condors flew down to join the fun. One of the new condors pushed the other one into a patch of melting snow, Sibley said; that bird turned on the tethered Topa Topa and "beat the shit out of it":

> [Our bird] managed to escape at one point, and raced full tilt to the end of the leash and surprise: the water-softened soil no longer held the stake and the young bird was off and soaring over the chaparral. It immediately went below the line of sight and [a colleague] lost it, a frantic search turned up no bird. When last seen the wooden stake was still trailing from the leash.

Sibley knew the condor was young because its head and underwing feathers were still black, but past that he knew nothing. He didn't know where the condor lived or who its parents were. He didn't know how old it was, or why it looked like such a zombie or whether anyone would ever see it again.

More frantic searching followed. More biologists joined the chase. For several days they again scoured the rugged countryside, looking for a bird that hardly anyone expected to find alive. Sibley finally found it when he smelled something awful emanating from under a big bush. Crawling underneath the bush he saw Topa Topa hanging upside down like a giant roast chicken in the window of a deli, its leg still fastened to the tether. The stake at the other end of the tether was caught in the branches of the bushes. Sibley climbed up and pulled it out, recalling that the condor "thanked me for finding him by taking a very painful chunk of flesh out of my arm."

Topa Topa lay on its side in a cage in the front seat of Sibley's pickup truck as he drove it down to the zoo. The tethered leg was clearly injured, but Sibley didn't know how badly. When he reached the zoo, he gingerly set the bird down in front of a bowl of food. Almost immediately, Sibley said, Topa Topa "stood up on its tiptoes and stretched like it was ten feet tall, walked across the floor, and gobbled up the food. The collective sigh of relief was amazing."

Topa Topa hasn't seen the wild since. He's been at the Los Angeles Zoo for thirty-seven years now, and there have been few dull moments. Attempts to put Topa Topa in a cage with other vultures failed when he bit the wattles off their necks; he also bit the lips of a llama in an adjoining cage. Topa Topa often flew across his pen and slammed his feet into the chests of human visitors, knocking them through open doors or onto their backsides. And whenever keepers tried to examine him he tried to eat their clothes. "Shoelaces were ripped loose and shoes were torn apart," wrote bird keeper Frank Todd in one of a series of reports he prepared for the zoo. "There was usually nothing left of the socks above the tops of the shoes, and on one occasion Sibley's pants were ripped open from pocket to cuff." When Topa Topa was moved to a cage with an artificial cliff, he proceeded to destroy it. When other animals were housed with him, he refused to let them eat. Keepers started refusing to even enter Topa Topa's cage for fear that he would hurt them.[3]

"If you didn't like having little pieces ripped out of you, he wasn't a good bird to be around," Todd said. "Especially if you showed the tiniest hint of fear. What you had to do was hold your ground and maybe even charge back at him. Then he respected you."

Topa Topa seemed to enjoy biting the hands that fed him. Todd said he remembers one biting incident above all the others. It took place in the early 1970s, when the U.S. Postal Service issued a set of endangered species stamps, one of which featured a picture of a condor in flight. "The Postal Service called us up to say they'd like to take a promotional picture of Topa Topa and his keeper standing next to the Postmaster General, and we said 'sure,'" he recalled. "So they came into the cage and I had my arm around old Topa and we're all ready to get it over with. As the photographer is ready to shoot, Topa sticks his beak up into my sleeve like he wants to play around. Then I felt him bite a big chunk out of my arm. Nobody else saw him do it, that's how fast it was. I sat there waiting to get my picture taken with all this blood gushing down my arm."

Todd said some of Topa Topa's other habits were much more endearing. When it rained, the bird raced back and forth across its pen trying to catch individual raindrops. Topa Topa also had a dance he only did when he wanted Todd in his pen. "It consisted of running about on the ground, flapping the wings, jumping into the air some three feet or more while rotating one hundred and eighty degrees, and then either running over to the door or jumping up and down from its perch."

There was one thing Topa Topa didn't do that caught Todd's attention: he never made any of the noises that most condors make all the time. Condors don't have voice boxes, but they can hiss like dragons if you get too close to them. Todd thought Topa Topa could have made these noises, too, but for unknown reasons he did not. "The only sounds I have ever heard from him were when he was a year or so old," Todd wrote, "and that was merely the passing of gas."

———

The Sespe Creek project skidded off the tracks in the summer of 1967, when the citizens of Ventura County voted not to fund it. The tax referendum that would have raised the money for construction only lost by thirty-six votes, but the hands-off forces were thrilled by what they saw as a trend-setting victory. "The conservation philosophy out of which the Sespe sanctuary had materialized was spreading like a leaven through our society," Ian McMillan wrote.[4]

McMillan and his allies failed to thank Fred Sibley for helping to nail this coffin shut. Since 1965, Sibley had been quietly demolishing the arguments used to support it, including the offbeat notion that a massive building project would improve the condor's life. "The Sespe Creek project can only be judged as being unacceptably detrimental to the condor's survival," he wrote at the end of a federal report on the impact the project would have had on the future of the giant birds. "The extinction of the condor would be an almost certain consequence."[5]

Sibley quit his job as a condor biologist after filing this report in

1969. It's often said that he was driven off by McMillan, but Sibley says that's not what happened.[6]

Sibley says he quit in the wake of the Santa Barbara oil spill, after a superior ordered him to stop talking to reporters. "I said something about the spill that didn't go over well in Washington," he said. "They made it clear that I would not do it again, and so I found another job."

The Santa Barbara spill of 1969 was the first spill to be nationally televised. Afterward, especially in California, environmental activism surged, and plans to dam the Sespe Creek no longer had a chance in hell.

CONTINGENCIES

It is not always possible to say whether a certain oil development, logging operation, public use facility, or new roadway will definitely impair the condor's chances for survival. . . . Field biology is at best an inexact science.

> —Sanford Wilbur, California Condor
> Recovery Plan, U.S. Fish and Wildlife
> Service, 1979

Richard Nixon saved the California condor on December 26, 1973, when he signed the modern version of the U.S. Endangered Species Act into law. Whether Mr. Nixon gave a damn about endangered species will ever be an open question. After all, he *had* tried to weaken the bill as it made its way through Congress. But those attempts had failed, and the president needed a law he could brag about. And so he signed the version of the ESA that is now one of the country's most loved and most hated environmental laws.

Compared to the declawed version of the ESA of 1967, the new Endangered Species Act was potentially smilodonian, in the sense that it had long, sharp teeth that could do lots of damage. This was the version of the ESA that made it a federal offense to "take" an endangered plant or animal, and that word "take" had been defined to cover a huge range of human actions. You could still shoot a grizzly bear if it kicked down the front door of your house, but you couldn't fill a ditch with a rare animal living in it unless the feds came out and gave you the okay. Regulators at the Environmental Protection Agency and the National Marine Fisheries Service were part of the picture, too: they had to examine and sign off on plans to build dams on rivers that might be used by rare kinds of salmon.

When plants and animals were added to the federal endangered species list, certain things were supposed to happen quickly. Federal biologists were legally bound to charge into the field and decide which parts of the animal's range were absolutely crucial, potentially crucial, and so on. Regulators had to draw bright lines around the so-called critical habitat. People in the field were to get to work on plans to "recover" rare species, which basically meant building up populations until the rare plant or animal could survive on its own.

The ESA gave the biologists sixty days to map these "critical" habitats, which often meant they had to guess at where the forage lands should be, or at where the most important breeding grounds were found. It was assumed that most of these new "critical habitats" would not be large.

But when it was announced that the California condor would be the first to have its territory marked, the stakes got much, much higher. This was a bird whose recent range included almost all of California, and whose not-so-recent range stretched from Canada well into Baja California. Which parts of that range were the essential feeding grounds? Were there any nests on private land? What sorts of developments would spell the condor's doom, if the bird was not already past saving?

There were lots of questions just like that one now. For example, could spreading suburbs be blamed for the condor's continuing slide? What about the spreading smog? What about the hunters? What about the loggers and miners and the nudist hikers who threw rocks at the birds to make them fly?

What exactly was the condor's problem anyway? When work on the first Endangered Species Recovery Plan began in 1973, that question hadn't been answered. For all the work done by Koford and Sibley, guesses at the causes of the bird's decline were still just guesses, and the leading guesses varied wildly. Ian and Eben McMillan blamed a widely used squirrel-killing poison called Compound 1080, even though they never found a trace of the poison in the carcass of a condor. Farmers who had carpet-bombed their crops with DDT were also viewed as suspects, even though DDT killed eggs of birds that ate live prey, and not the scavenging condors.

The job of turning arguments into a legally defensible document fell to Sandy Wilbur, a genial and quietly relentless biologist labeled as "God's condor man."

═══

Sandy Wilbur says he used to be a scientist who believed in evolution. Then one day in the 1950s, Wilbur read a book by C. S. Lewis and decided to be born again. From that point on he became a scientist *and* a creationist: not the kind of detail I'd normally note, but in Wilbur's case it made a difference. Not because it influenced his work on the first recovery plan; Wilbur's beliefs gave him a reason to do that work.

A story on the Internet started me down this path of inquiry. Somebody had e-mailed me a reprint from a magazine called *World Wide Challenge,* published by the Campus Crusade for Christ. "Sanford 'Sandy' Wilbur is God's condor man," read the first line of this story. "He looks upon his work as God's call to preserve this species."

The author of this article put Wilbur in the middle of a fight over what a condor is. Is it a bird that "must be moved aside to allow for 'economic growth'?" How about a relic bird that "serves no useful purpose" and is too expensive to maintain? Could condors be "haunting symbols of life itself?"

In the middle of this battle stands God's condor man. He offers a different but surprisingly simple reason for his motivation and for his efforts to save the bird: "God didn't make any mistakes in what he put here to begin with," says Sandy. "He made sure that in the Great Flood everything was preserved. I don't see any reason why we should do less. I see helping to save condors as doing the work of the Lord."[1]

I'd already set up an interview with Wilbur when I read the piece. When we got started, I asked him if he'd mind explaining why his job was "the work of the Lord." Wouldn't it have been more reverential to leave the condors to their fate? Did he ever feel like he was "playing God" when he tried to change the condor's world? Did he mind me asking?

"No," he said, "not at all." Wilbur then explained that he'd been fascinated by the condor since he was growing up in Oakland. When Koford's work was published, he was only twelve, but he read all about it in the papers. "I was fascinated by the condor's rarity and its size," Wilbur said. "And by the fact that it had once lived in my backyard." Years later, when Wilbur was a federal biologist in Georgia, he heard that Sibley had quit, and applied for Sibley's job. "We prayed about it as a family," Wilbur explained. "It was a special opportunity and a very special trust."

Wilbur thought saving condors was a good job for a serious Christian. First of all, he'd be working for a cause with no particular significance beyond "taking care of something God made."

"Saving the condor didn't really benefit anybody," Wilbur said. "This was one of those rare cases where losing a species wouldn't be a great blow to any ecosystem, since there weren't enough of the birds to make them an important scavenger." Wilbur said he also felt a special need to "develop a process for saving condors that emphasized cooperation, integrity, and credibility. That way we could set some standards for other preservation efforts. It was an extra bonus that my particular charge was not an obscure little-endangered spider running around a sand dune somewhere, but was a giant, majestic, unique, folklorish and controversial bird."

Wilbur felt "shocked and thrilled" when he was picked to replace Fred Sibley in the fall of 1969, even though he knew if he stuck around long enough, he might see the birds become extinct. Few endangered species had a range that covered half the state of California, or such a long and ominous list of potential threats to its existence, or such a short and inconclusive list of *proven* threats to its existence. Making matters worse was the glacial rate at which the birds replaced themselves.

"If you did to a duck what we did to the condor you'd have about as many ducks today as we [had] in 1800," he said. "The condor, on the other hand, was essentially a species with no built-in mortality factors except old age and the occasional accident. With a reproductive cut that even under pristine conditions could do little more than ensure that the breeding pairs replaced themselves sometime during their lifetimes, each loss to shooting, to poison, to being caught in a trap, to flying into a power line, to having an egg taken was enough to tip the scales against the species."

Wilbur tried to slow the condor's slide for most of the 1970s. He wasn't able to do it with the tools he had at hand. Activists such as David Brower got mad when Wilbur challenged the McMillan brothers, dismissing as hearsay the notion that the birds were being killed by a widely used squirrel poison. The activists got madder still when Wilbur said the condors might be having trouble finding

food. Grazing lands were shrinking and ranchers seemed less inclined to leave cattle carcasses where they fell. As these ranchers
hauled condor food off to dumps and incinerators, the birds themselves were forced to spend more time finding food and less time
eating it. Wilbur also thought the birds were laying fewer eggs and
bringing less food back to the caves. Fewer fledgling condors
seemed to take wing each fall, and that was a problem. Breeding
birds were not replacing themselves. The population was aging.

By the time Wilbur sat down to write the first official California
Condor Recovery plan in the middle of the 1970s, a reproductive
crisis appeared to be at hand. Old-fashioned efforts to preserve the
bird were failing, he'd decided. Soon it would be time to take some
much more drastic steps.

Wilbur defined these drastic steps in a Contingency Plan attached to the recovery plan, which was then sent out for public
comment. Hands-off activists were pleased by Wilbur's calls for restrictions on hunting, logging, mining, road building, and drilling,
and by the urgency with which he insisted that the government buy
private lands surrounding roost sites. Wilbur also emphasized the
need to learn much more about the possible impacts of common
pesticides and poisons.[2]

But the activists were outraged by the "last-ditch" actions out-
lined in Wilbur's attachment. If all else failed, every condor left in
the world would be trapped and placed in holding pens, where they
would be handled and bled. Nine of these birds would be shipped
to "captive propagation facilities," one of which would be in San
Diego. Some of the other birds would be released with experimen-
tal radio tracking devices bolted onto their tails and big numbered
ID tags attached to their wings. Field biologists would track and
plot the movements of the bugged birds, learning crucial things
about the way condors lived. Fights over whether they were finding

enough food would be settled once and for all. Scientists would not have to guess at whether unknown condor X was old or young or male or female.[3]

Some thought it was ironic that "God's condor man" would be the author of a plan that would essentially play God with the condors. Wilbur thought the critics had it backward, though. People who refused to do what it took to save the bird were the ones who would be playing God.

Some of the most pointed attacks on Wilbur's plan came from Carl Koford himself. Koford hadn't studied condors since the late 1940s, but that didn't seem to give him pause. At hearings and in letters to elected officials, he called the Wilbur plan a dangerous experiment that was likely to destroy the species. This was a position Koford shared with David Brower, and his friend Paul Ehrlich of Stanford University.

Ehrlich was the author of *The Population Bomb*, which warned that runaway population growth was wrecking the planet. He was also a regular guest on the *Tonight Show Starring Johnny Carson*, and he stuck with Koford to the end. In a letter to the Secretary of the Interior in 1980, Ehrlich denounced the contingency plan as a terrible precedent, a waste of money. "The projected expenditures for trapping, marking and putting radio transmitters on most of the remaining condors amounts to nothing more than unconscionable and dangerous harassment," he wrote. "Nothing of general scientific value could come from such observations that could not come from studies of birds not threatened with extinction."

Koford died of cancer in 1979. After a memorial service on the U.C.-Berkeley campus, Eben McMillan drove the great man's ashes to his ranch in central California, where he threw them into the wind.

Not long afterward, the federal government and the state of California announced that it was time to put Sandy Wilbur's rescue plan into effect. In doing so, the regulators turned their backs on some remarkably eloquent cries of dissent from hands-off activists such as Brower, who argued that at this rate it would not be very long before travelers in Southern California would be greeted by signs along the highway that read LOS ANGELES–NEXT 250 EXITS. In an essay entitled "The Condor and a Sense of Place," Brower wondered whether the city to the south would ever stop expanding:

> Must Paul Bunyan move to California, go into real estate and ride a giant leaping frog, leaving colossal subdivision plans at landing? . . . Count on Japan's present population as the model for California's Year 2080, and China's for the United States as a whole; after all, we are talking about roughly the same respective areas. Such a grim future, with its coalescing cities and suburbs, will have far too little room for people and no room at all for condors.

Koford's written objections to the last-ditch plans were read aloud at public hearings held before key votes. "If condors are not surviving well in the wild," he had written in an essay, "should we expect released cage-raised birds to do better? In the wild they must forage skillfully, know the landscape and air currents, seek appropriate shelter at night and in storms, cope with aggressive eagles and compete with established condors. All poultry breeders know the difficulties of adding to an established flock a new bird: it is generally rejected and killed. Must we further dilute the natural scene by mishandling the birds and injecting cage-raised stock into condor society?" Koford ends this essay by insisting that a "cage-raised" condor could never be more than a "partial replicate" of the real thing, adding that "If we cannot preserve condors wild through understanding their environmental relations, we have already lost the battle."

Words like those did a lot to influence public opinion, but by the late seventies, they didn't seem to carry much weight with relevant federal and state officials. This was true in part because the hands-off point of view had been officially considered and rejected by a panel of prominent wildlife experts called in to do a "scientific audit" of the recovery plan.

"The existence of the California condor depends on conscientious human intervention," read the panel's final report. "This will always be so. The only reasonable hope for achieving a large population of condors in the wild is captive propagation."[4]

When the scientific panel dismissed arguments put forward by Brower and his allies as "vacuous," Wilbur said he felt both vindicated and enormously relieved. And yet, not long afterward, Wilbur said, he was unexpectedly reassigned to a job in Sacramento. A new team of biologists was coming in to carry out the last-ditch rescue plan, and they didn't want Wilbur's input.

The condor known as AC-9 or Igor watched trapper Pete Bloom catch several of the last wild condors from the top of a nearby tree. Igor himself was trapped on Easter Sunday, 1987. (Photo by Dave Clendenen; courtesy of Peter Bloom)

In 1986 and 1987, the last free-flying California condors were trapped by biologists like Peter Bloom (pictured) of the National Audubon Society and Dave Clendenen of the U.S. Fish and Wildlife Service. (Peter Bloom)

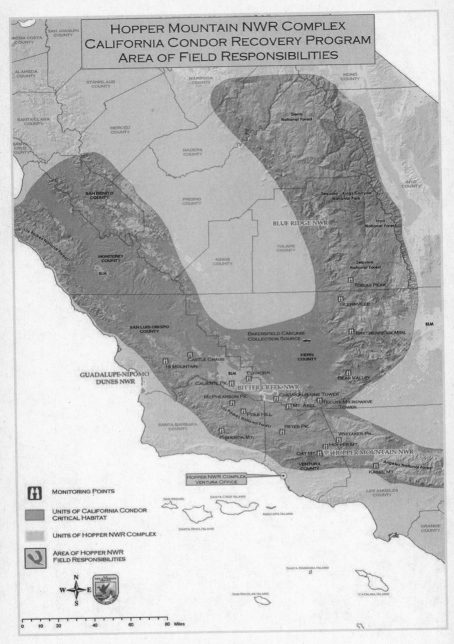

After World War II, the range of the condor followed
this wishbone-shaped set of mountains.
(Courtesy of U.S. Fish and Wildlife Service)

When creatures like the saber-toothed cat and the mastodon were alive, the California condor ranged across large parts of North America. (Knight Mural of Pleistocene Life, Rancho La Brea Tar Pits #4948 courtesy of the American Museum of Natural History)

Indian tribes in California revered the birds and left paintings of condors on rocks. (Courtesy of U.S. Fish and Wildlife Service)

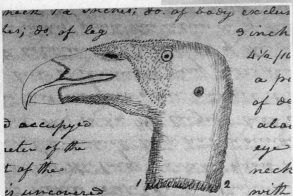

"I believed this to be the largest bird in North America," wrote Lewis in his journal in 1806; Clark wrote the same thing and added a rough sketch of the bird's head. (Courtesy of the American Philosophical Society)

Hundreds of years after Lewis and Clark explored the Columbia River, condors were returned to the Vermillion cliffs and the Grand Canyon. (Christie Van Cleve)

In 1840, John James Audubon immortalized the "California Vulture" in "The Birds of America." (Courtesy of Haley and Steele)

By the early 1900s, hunters and egg collectors had all but wiped the species out. (Photo by R. Corado; courtesy of the Western Foundation of Vertebrate Zoology)

Joseph P. Grinnell, a legendary naturalist, was among the first to describe the bird as a "symbol of our lessening wilderness." (Courtesy of the Bankroft Library, University of California, Berkeley)

In 1939, Carl Koford was the first to study the behavior of wild condors and urged hunters, loggers, hikers, photographers, and scientists to stay away from the species. (Courtesy of Rolf R. Koford)

At times, Koford and some of his friends handled condors in their nest caves. Koford later denounced the "hands-on" approach as a threat to the future of the species (Ed Harrison, above). (Photo by Carl Koford; courtesy of Lloyd Kiff)

In the desperate 1980s, when ravens were seen breaking condor eggs and eating the contents, biologists like Rob Ramey shot at birds that approached the nest caves. (Courtesy of Rob Roy Ramey II)

By the 1980s, almost all of the remaining wild condors had been trapped at least once, and most had numbered ID tags and radio transmitters hanging from the front of their wings. (Christie Van Cleve)

In 1987, the last wild condor was captured and taken to the San Diego Wild Animal Park. (Courtesy of the Zoological Society of San Diego)

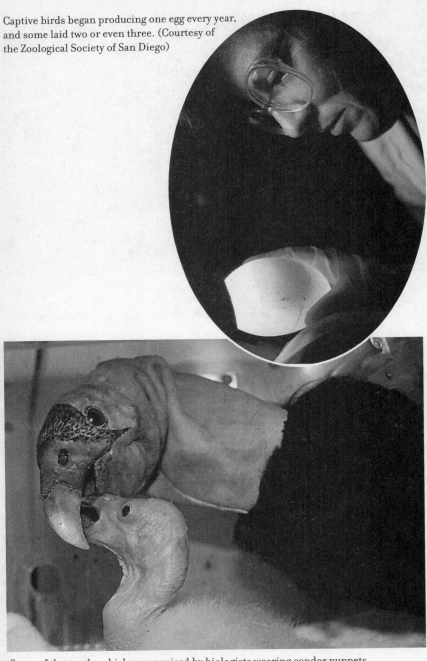

Captive birds began producing one egg every year, and some laid two or even three. (Courtesy of the Zoological Society of San Diego)

Some of the condor chicks were raised by biologists wearing condor puppets on their hands. (Courtesy of the Zoological Society of San Diego)

Even the notoriously antisocial condor known as Topa Topa fathered an egg. (Anthony Prieto)

Residents of southern Utah and northern Arizona said they didn't want the condor around. (Christie Van Cleve)

In 1996 the birds were released near the Grand Canyon, where they sometimes buzzed the crowds on the South Rim.

Condors often soar for hundreds of miles in a single day. (Photo by Christie Van Cleve)

Some birds raised in zoos have had trouble learning to be wild, turning up at Burger Barns, airport runways, and on the decks of homes built on the sides of mountains. (Photo by Denise Stockton; courtesy of U.S. Fish and Wildlife Service)

Condors are threatened by lead fragments found in the carcasses abandoned by hunters—and by the nuts and bottle caps that sometimes show up when the birds are X-rayed. (Photo courtesy of the Los Angeles Zoo)

In the Hopper Mountain Condor Refuge in California, birds are regularly trapped and tested for lead. (Photo courtesy of U.S. Fish and Wildlife Service)

Condors that fail to adjust to the wild are flown to the zoos
for treatment. Most are eventually returned to the wild.
(Photo courtesy of U.S. Fish and Wildlife Service)

The condor that was known as AC-9 or Igor was returned to the wild in
the fall of 2001 with a new set of numbers on its wings. Condor biologists
who track his movements say he seems to be thriving. (Anthony Prieto)

The matriarch, also known as
AC-8, free in better days.
(Anthony Prieto)

Last year, after having a tumor removed and surviving a severe case of lead
poisoning, the matriarch was shot dead by a pig hunter who said he didn't
know what condors were. (Courtesy of U.S. Fish and Wildlife Service)

eleven

ENDGAME

===

(i) The California Condor is rapidly declining to extinction. The population's future is particularly precarious because the remaining individuals, certainly not more than forty and possibly fewer than thirty, are concentrated in a relatively small area that is subject to intensive development.

(ii) The existence of the California Condor depends on conscientious human intervention. It cannot survive without careful management and protection of its environment. This will always be so.

> —Report of the Advisory Board on the
> California Condor, submitted to the
> National Audubon Society and the
> American Ornithologist's Union on
> May 28, 1978

Think of it as protective custody.

> —Arthur Risser, San Diego Zoological
> Society, 1987

A scientific SWAT team rolled into the Sespe in the summer of 1980, ready to do whatever it took to save the condor from extinction. Fights over whether the birds should even be approached by humans were now officially beside the point. The condor's only hope was what a panel of prominent scientists had described as a "long-term large-scale program involving a greatly increased research effort, immediate steps to identify and con-

serve vast areas of suitable condor habitat, *and* captive propagation." Activists such as David Brower and the McMillans still
seethed when people mentioned the zoos, but a panel of esteemed
ornithologists had swept those objections off the table by arguing
in no uncertain terms that the time to take a number of "drastic
steps" had come. Captive breeding wouldn't be enough if the
birds not taken to the zoos were left alone. According to the panel
those birds had to be trapped, "individually marked and fitted
with radio transmitters, provided it proves possible to do so in a
reasonable time and without undue trauma to the birds." The tags
and transmitters were supposed to make it easier to follow the
condors around, which in turn would help researchers study "sea-
sonal distribution, daily foraging range, habitat use, population
size, and reproduction by free-living birds." All of these things
could be done in other ways, but not as quickly, and time was
now of the essence.[1]

No endangered species team had ever taken on a job this big
and complicated. On the other hand, there'd never been a team
that seemed to be so well prepared. Congress had not only
agreed to spend millions to support the last-ditch program, it
had also agreed to formally deputize the National Audubon Society by allowing it to share control of the new program. The
man the society picked to represent it was John Ogden, a highly
regarded ornithologist based in the Florida Everglades. The federal half of the director's chair was taken by Noel Snyder, the
leader of the Condor Recovery Program in the 1980s and a raptor expert with the U.S. Fish and Wildlife Service. Advocates of
the hands-off school of condor management thought Audubon
had traded its soul, but other environmental groups were prais-
ing the society's willingness to put its staff, and a lot of money,
where its mouth was.

Ogden and Snyder helped save the condor from extinction, but the price they paid was very high. Terrible mistakes were made in the field while they were in control. And every time something went wrong, they were blamed.

In the end, the pressure split the field teams they worked with into warring factions. Plans to keep some of the remaining condors in the wild were eventually abandoned, and at one point the Audubon Society sued to stop the trapping.

But when Snyder and Ogden first drove their trucks into the sanctuary, bitterness and desperation were not part of the plan. Instead of trapping birds that might be followed to the zoo by angry crowds of protesters, they'd decided to start with something quiet and safe. Plans were laid to enter two nest caves and handle two chicks, taking various measurements and running various health tests. Both operations would be short and sweet, and both would be filmed and photographed. The footage would be used to help drum up support for their work.

That was the plan. Implementation began the next morning when the two men led a field team to the edge of a sandstone cliff eighty feet above an active nest site. Two adult condors were inside, tending to the chick that would later be known as Igor. The parent birds had long been known as a bitterly unhappy couple, always bickering and pushing each other out of the way, and the bickering had gotten worse since Igor's hatching. Birds such as this have been known to attack approaching humans, but this pair wasn't so tough, flying off the minute they saw Snyder heading down to their cave, followed by a young biologist and climber named Bill Lehman.

Snyder went in ahead of Lehman, quickly cornering Igor, who was then a fuzzy feathered blob with tiny legs and a tiny beak. Igor didn't struggle when the vet picked him up and measured his stubby black wings, or when the vet put the bird into a horse's feed bag so he could weigh it. "He was calm and curious and totally relaxed," as Lehman remembered. "It was as if he'd been expecting us."

The vet took a few more measurements and a sample of Igor's blood. In the future, he planned to do the same to other condor chicks, taking the same measurements and running the same health checks. That would help build on the information gathered by Koford, Sibley, and Wilbur, thickening the baseline and unraveling the answers to especially vexing questions: Why do some of the breeding pairs fail to make eggs? How do condors find their food? How do the birds protect their eggs from bears, coyotes, eagles, and ravens for starters, and from a host of smaller predators after that? Are the condors starving or are they being poisoned? Are the numbers going up or down?

"We needed information," Snyder said. "Everybody had a theory then but nobody had the database they needed to support their point of view."

Igor's checkup lasted ten minutes. As the researchers climbed back up the cliff, the parents returned and went right back to bickering, acting as if they'd never left. Snyder and Ogden were encouraged by this, and so they agreed to go ahead with the plan to examine a second chick in the morning. Then, they'd catch a flight to Africa to study the habits of some of its most condorlike vultures.

"We were off to a great start," said Snyder. "There wasn't any reason to think it would change." He said he and Ogden were in high spirits when they reached another cliff the next morning. Neither man could see the cave below this cliff, or the chick in the back of it. But they knew the cliff was steeper and more treacherous than the one they'd climbed the day before, so they decided to play it safe by sending down a smaller team. Since Ogden, Snyder, and the still photographer weren't skilled climbers, they decided to remain on top of the hill above the nest site. But Bill Lehman and the cameraman were both professional climbers, and they both wanted to go down.

Lehman dropped over the edge first, moving swiftly and easily down to the cave entrance.

"I entered the cave entrance and saw the chick back away and hiss a few times. It was much bigger than the other one, and a bit more frantic. Took me a while to corner it and pick it up, but then it squirmed and got away."

By then, the cameraman was behind him. With the camera rolling over his shoulder, Lehman maneuvered the chick back into the corner and caught it a second time. It was still squirming, and he was forced to hold it tighter than he wanted to. After a few seconds, he managed to squat down and stick the bird between his knees. He reached into his backpack and took out a caliper.

Jan Hamber, a condor expert with the Santa Barbara Museum of Natural History, was watching the scene through a sighting scope on a distant cliff. "I could see that Bill was having trouble," she said. "The bird was putting up a fight and things were taking a long time. When he got the measurements done, he tried to put the bird into a feed bag so he could weigh it, but that didn't work."

The chick was too big. Lehman thought it might have fit into the bag if he'd pushed it a little bit but he didn't want to do that. So he put the chick back between his knees, emptied out his knapsack, and put the chick in there. He put it on the scale, took the bird out, and weighed the empty backpack.

It was at about this time that Lehman started feeling queasy. He'd had his hands on this chick for a long time, and something about the way it was acting worried him. But he didn't know what it was and he kept going. He was in the homestretch now.

"I was a good soldier," he told me later. "I was going to get the job done. I was taking the very last measurement when the chick started shaking."

The cameraman turned his camera off and yelled at Lehman— "Something's wrong with the bird. Put it down! Put it down!" Lehman put it down. It wobbled a bit, then fell over.

Lehman can't remember whether he had a walkie-talkie with him in that cave. If he did, he can't remember using it. "When the chick started having problems I walked out to the front and yelled, 'Hey! We've got a problem down here!' Or something like that. They yelled back and told me to sprinkle some water on the bird, so I did. But by then it was too late."

The chick had stopped breathing. Lehman stared down at it, knowing it was dead. After some more yelling, he picked up the carcass and put it in the backpack. He tied the backpack to a rope thrown down from the top of the cliff, and stood back to watch the bird ascend. Then he climbed up after it, moving very slowly.

"That was the longest climb of my life," Lehman says. "It seemed like it took six months. When I got to the top Noel was right there, and right away he asked me to show him how I held the bird. When I did, he said, 'Okay. Don't worry. It's not your fault. This is my responsibility.' He didn't have to say that, but he did, and I will always be grateful."

Statistically this was a flesh wound. But Snyder knew that wasn't the point. "The condor had been so deified by the wilderness people that it was going to be like killing Jesus Christ to lose this bird."

Snyder was right about the activists—they were in a state of rage before the carcass of the chick was off the mountain. Dave Phillips of Friends of the Earth got an early call from a friend at the Sacramento office of Fish and Wildlife. He turned around and called David Brower. Brower talked to other environmentalists and a lot of reporters and elected officials. Then he sent a telegram to Cecil Andrus, a friend who was also the Secretary of the Interior, urging him to fire the Snyder and Ogden teams and abandon the last-ditch recovery plan. Other activists urged the state of California to do the same thing. Some made a special point of questioning the role of the National Audubon Society in an episode that was variously described as an act of "stupidity in its most basic form," a "breach of responsibility," and a "glaring dereliction of trust." All of those phrases appeared in a letter that was written by Eben McMillan; it urged the California governor, Edmund Brown, to reverse the state's decision to allow the National Audubon Society to function as a quasi-judicial agency. "The National Audubon Society has no place in setting policy nor of being responsible for the welfare found within the confines of our own state lands," he wrote. "Audubon is a private agency in no way beholden to the demands of our people."[2]

Brower and Phillips asked to see the film of the disaster. Snyder says he sent it without delay. Hands-off activists showed the film to local groups of bird-watchers, criticizing Lehman's every move. Walter Cronkite mentioned the death of the chick on the *CBS Evening News*, and Dave Phillips did a frame-by-frame analysis for National Public Radio.[3]

Spokesmen for the new research team said the bird had been properly handled. A necropsy report from the San Diego Zoo listed the cause of death as a heart attack triggered in part by "stress," adding that the heart of this particular chick was slightly larger than normal. The implication was that the chick may have been unusually fragile.

Brower and his colleagues jumped all over that claim, charging that Lehman, an obvious novice, should never have been asked to single-handedly restrain and measure an oversize condor chick in a cave where he couldn't be seen or heard by his superiors. Brower added that he was appalled by the claim that only Lehman and the photographer had the climbing skills needed to rappel down to the cave: "Having helped teach some ten thousand Army men how to rappel safely [and a Secretary of the Interior, too], I know the team should have and could have learned to get down to the nest themselves."[4]

Brower urged Lehman to quit the recovery team and join the hands-off faction. "As the person in whose hands the condor chick died, and as a man who is obviously deeply distressed by the tragedy, you could wield enormous influence," he wrote in an open letter that urged Lehman to call for a return to more "Kofordian" condor research, in which "ecological understanding" would be reached by "careful, patient observation."

Lehman took the offer to defect as an insult. First of all, he thought it had a patronizing tone. Second, he hated Brower's habit of sending copies of his letters to the press. Most important, Lehman didn't think Brower knew what he was talking about. In his view, the time for "careful, patient observation" was officially over.

While these kinds of interchanges were under way, Snyder was offering to quit. He said his superiors in Washington rejected his offers: "They told me it would look like an admission of wrongdoing if I resigned." Snyder replied that his offer to quit was supposed to be just that. But he understood the logic and so he stayed on, working with Odgen. Both men knew the hands-on program would soon be suspended and perhaps abandoned completely. Snyder said it never crossed his mind that this fiasco might turn out to be a lucky break. But in a way that's just what happened.

━━━━━

Weird things transpire when you sit behind a fake wall of leaves in the middle of nowhere watching condors all day long for weeks on end. First, you forget what time it is. Then you lose track of the days. You watch a group of male tarantulas march en masse in pursuit of some females. You see a rattlesnake slither past you to the edge of a puddle, where it flicks its tongue in and out of the water like a strong but flickering flame. After that you might see a makeshift balloon with something hanging underneath fly by: raising your binoculars, you see what firefighters call "an arsonist's balloon," a flaming can of something that resembles Sterno hanging from a paper bag.

Then, if you're lucky, you'll point your binoculars at the nest cave again, just as a tiny beak busts through the top of a big bluish egg. You might see the parent birds push the egg off a cliff by accident. You might see a golden eagle strike like a flash of lightning. Parent condors might play with their young out on the ledges, watching you watch them watch you.

"We'd called it 'living on condor time,'" said biologist Helen Snyder. "If you wait long enough, you grow invisible to the world around you."

The California Condor Recovery team did a huge amount of living on condor time in the early 1980s after the chick died in the arms of Bill Lehman. If the team had kept its trapping permits, Noel and Helen Snyder might not have been the first to see a condor egg hatch in the wild, or the first to see a raven slurp the yolk out of a condor egg it had just punctured with its beak. Most important, they might have missed the chance to be the first to see a pair of condors make two eggs in one breeding season, laying the second after the first was accidentally lost. Ornithologists call this process "double clutching," but they weren't sure the condors did it. Noel Snyder said he felt euphoric when it happened.[5]

"It proved that we could take a set of eggs to the zoo without making the wild flock smaller," he said. "It meant we had a chance to double the number of condors hatched each year. Of course we were even more amazed when we started seeing triple clutching. That happened several times."

Snyder's other breakthrough came when he figured out how to tell the condors apart. This was supposed to be impossible to do from a distance, he said. This was true because male and female condors have the same markings, and because the birds don't vary all that much in size. It also explained why previous attempts to count the birds had produced such different results—if a condor flies out of sight and then back again, how does an observer know it's not two different birds? Trapping and marking every last condor left was one solution, but at the moment, that was not allowed.

Snyder said the answer came to him when ornithologist Eric Johnson of California Polytechnical State University at San Luis Obispo brought a group of students up into the mountains to photograph the birds. When he saw their pictures, he noticed that some birds had broken feathers on their wings; others had feathers that were merely bent or molting. If someone were to organize a mass photography session in the condor's rangelands, those photographs could be studied and sorted into piles, one pile per bird. If Snyder were to count the piles he would know how many birds were left. If there were a lot of them, captive breeding programs wouldn't seem quite so essential. If there were only a few, it would be the other way around.

Snyder and Johnson started handing out cameras to every reliable observer they knew. Not long afterward, towering stacks of condor pictures filled Snyder's office. Some of the pictures showed birds with telltale scars on their heads. All the other clues were in the wings. Photographs taken in the course of a day were laid out on a map drawn on the floor of Snyder's office, so that Snyder and

Johnson could trace the flight paths of individual birds. After several weeks of sorting, the two men concluded that as of the summer of 1982, there were no more than twenty-four and no fewer than twenty-one California condors left on Earth.

The fact that the birds had finally been counted was what Snyder called the good news. The count itself confirmed his fears. When Snyder staged a second count in 1983, the total was even more depressing: that time he ended up with twenty-two piles of photographs. "These data . . . indicate a continuing catastrophic decline of the species," he and Eric Johnson later wrote. "In the absence of intense conservation measures, extinction of the wild population can be expected in 10–20 years."

While the photographic census was under way, the field teams kept on watching the birds, which was not an easy thing to do. "Everybody burned out eventually, but some people seemed to have a gift," said Helen Snyder. "Some people could sit and watch nonstop for weeks, and some went nuts after just a few days. Most people lasted about five days before they started missing things. After that it turned into, 'Gee I'd like a beer,' or, 'I wonder what my girlfriend is doing,' or, 'I could be in Santa Barbara right now.'"

One researcher never seemed to want to leave his home in the wild. He was Jon Schmitt, a line artist, careful note-taker, and accomplished taxidermist. "Jon would sit there all day every day for a month at a time," said Helen Snyder. "And none of his field notes ever missed a beat. I don't know how he did it, but it never seemed to bother him. He's the guy who saw the golden eagle try to kill the chick at the entrance to the cave."

Schmitt told me he'd be happy to describe the attack once he reviewed his field notes. He didn't want to mix it up with any of the dozens of other condor-eagle episodes he'd seen while sitting in the blinds. Some of these encounters were halfhearted eagle attacks designed to push condors out of choice thermals. Other attacks were so relentless and aggressive that they ended with the condor lying on the ground or hiding in a cave.

Schmitt said he knew there was an eagle on the way when he saw descending condors looking backward over their shoulders. Anything below a diving golden eagle is in trouble, but Schmitt said the condors had their defenses up. Once he watched a condor fleeing from an eagle that was shooting down out of the clouds, gradually increasing its angle of descent. "By the time the eagle and the condor were in the same optical field of view they were falling almost vertically," he said. "Both had their wings drawn close in and their flight feathers were swept back."

Schmitt heard "a loud shrill tearing sound" as the birds shot past, even though they were three hundred yards away. "The condor is falling hard and fast, but the eagle is so much faster. It closes the distance swiftly, and just when I thought contact had been made, the condor deftly rolls over, briefly flying upside down." The anchor-shaped eagle sheared harmlessly past the suddenly inverted vulture; almost instantly, it vaulted up, "like it had bounced off of some kind of invisible surface." Golden eagles like to stay above their prey, and this one wasn't taking any chances. Schmitt said it rose almost vertically, several hundred feet up into the air in a matter of seconds.

The eagle thought the condor would attempt to keep pace, but by then the condor was gone. Schmitt said he saw it blasting forward just above the ground, entering a forest of fir trees that tapered out near a popular roost site. Schmitt couldn't see the roost site, but he figured that that's where the condor ended up. Even golden eagles are hardpressed to cope with several condors at once. The eagle flew over the horizon toward the roost site, Schmitt said, but a few minutes later it flew away, chased by heckling ravens.

Eagles also tried to get at fledglings. Schmitt saw this on two occasions, one of which began when a golden eagle dove on a squirrel that managed to escape. Schmitt was in a blind about a quarter-mile off, looking at the back of the eagle, when he saw it lean horizontally forward with its wings pressed slightly outward. "It was weaving its head from side to side, triangulating the distance to its next intended victim. I was looking over the eagle's shoulder at a helpless condor chick, which for some imbecilic reason had chosen this moment to walk out to the front of a cave on the far side of a gorge." The eagle took off and headed straight for the chick, which never seemed to see it coming. But at the last second, the eagle was knocked off course by a dive-bombing parent condor.

Schmitt was in the same blind when an eagle crashed down into a condor that was feeding a chick. He said the chick bounced into the cave while the adult turned around to face the eagle, which was lying on its back with its talons pointed up into the air. "Wings flailing and audibly slapping and scraping rock, the condor is standing on the eagle, furiously tearing at the eagle's breast and throwing clumps of feathers to the side; the eagle is ripping at the condor with its long legs and huge talons." When the eagle broke free and attacked again, the birds pushed each other off the cliff. Both took flight before hitting the ground. Schmitt said the eagle flew away.

Everyone on the field staff had stories to tell. Some had found Indian burial sites. Some had discovered very old guns. Many saw fighter jets from Vandenberg Air Force Base chasing low-flying missiles up the canyons. "Nonflapping entities" were not supposed to drop lower than three thousand feet above the Sespe, but that rule was frequently ignored.

The speed with which the missiles and the planes roared through appeared to freak the condors out, as did the sonic booms they left behind as calling cards. Condor watchers saw the birds leap up off their eggs when they heard these booms. Every now and then a soaring condor would be rocked by the winds bouncing off a passing fighter.

Snyder and his team were out in force when a space shuttle passed through the refuge on its way to Edwards Air Force Base on July 4, 1982. The boom that trailed those rockets made the roar of the missiles and the fighter planes sound like a mild summer breeze. Snyder said his colleagues saw a parent bird explode out of a nest cave as the shuttle passed, rushing forward in a way that made it look a little like a giant feathered cannonball.

Snyder and Ogden requested that the Navy ask its pilots to fly higher. They also urged NASA to bring their space shuttles down in Florida when the condors were tending to their eggs. To prove the need for these actions, Snyder and Ogden sent a package of photographs. One of them was taken just before an unarmed cruise missile crashed into the ground near the condor's breeding grounds; Jon Schmitt was the one who saw it happen. He'd learned to identify different kinds of missiles when he saw one waver and angle down in the spring of 1983. "I saw a puff of smoke before it fell into a canyon," Schmitt said. "At that point it was out of sight." Schmitt said the escort jets sailed forward over the tops of the mountains before turning back toward the canyon adjacent to the one that held the wreckage. Apparently unaware that they were looking in the wrong place, the fighter jets circled briefly and then left the area. Schmitt said he called the Forest Service on his walkie-talkie then, asking them to tell the Air Force where the missile was. A few minutes later the Forest Service passed a pair of messages back: First, the Air Force was aware of an "incident" that may or may not have involved a missile. Second, it did not need his help.

Shortly after that, a huge black military helicopter thundered forward, buzzing what may have been an occupied nest cave on its way up the wrong valley. Schmitt called the Forest Service a second time, asking that his message again be relayed to the Air Force. The Forest Service passed back a less polite version of the last set of messages. The helicopter looked around and flew away. Later, the Forest Service called back to ask Schmitt to repeat his earlier directions. The military helicopter came back the next day to remove the wreckage of the drone.

By the time Schmitt saw the crash of the cruise missile, the contingency team had its trapping permits back. Snyder's team was taking eggs from breeding pairs and giving them to the zoos. Ogden's team was laying out carcasses and chasing after condors that had radio tracking devices bolted to their wings. Snyder himself was following the tagged birds around in an airplane, picking up data points that he described as "nothing short of spectacular." Ogden said he watched the condors shadowing cattle herds during the calving season and then doing the same to deer hunters for the hunting season. He saw them soar on unseen winds for hundreds of miles at a time. He saw them stay away from logging operations on specific peaks, and noticed that some of the most traveled routes crossed mostly private land.

Ogden and Snyder were now fighting on a more or less constant basis. By some accounts they fought over control of the program, but by others they just didn't get along. The precedent-setting partnership between the U.S. Fish and Wildlife Service and the National Audubon Society was splitting into warring factions, with Ogden and the radio trackers on one side and Snyder and the field teams on the other. The split became a literal one when Snyder started running his half of the recovery team out of his home in Ojai, all but abandoning the Ventura office space he shared with Ogden.

For a time the only thing that kept these fights in check was the crushing load of fieldwork being done by each of the field teams. Ogden's crew was busy dragging carcasses around and chasing after condors wearing radio tags. Snyder's half worked so ridiculously hard out in the field that it's a wonder they're all still alive.

They called themselves "the zombie patrol," because that's what they often looked like when they staggered toward the condor nest caves—filthy, smelly, bleeding, starving, stiff, and utterly exhausted. Noel Snyder marched in the front of his group, carrying a pair of antiseptic gloves and a black padded suitcase that seemed to double as a good luck talisman: in the 1960s, other field biologists had used this case to carry whooping crane eggs off to captive breeding centers. Since condor eggs were roughly the same size as crane eggs, Snyder had the case sent to him when the permits came through that allowed him to take condor eggs. "It had thermometers sticking out of the top so we could make sure the eggs were warm enough," he said. "We were out there carrying this suitcase through the brush, which must have looked very strange."

Snyder was allowed to bring the lucky black suitcase out again in 1982, after he watched that pair of condors double-clutch to produce a second egg. Late in the summer of 1983, the field teams staked out every active nest they knew about. When the breeders laid their first set of eggs, the zombie patrol moved in, fording raging rivers, climbing vertical cliffs, cutting trails with chain saws, and clearing helipads in the middle of the night.

As soon as the team had the staging area all set up, Snyder and the pickup crew would climb up to the nest cave. When they got there, one of the researchers would tell a long, sick joke, in hopes that the sound of a human voice would lure the condors off the eggs and out of the caves. The joke was always told in a low voice, so as not to startle the birds: panicked condors might have crushed an egg on the way out.

The joke is too sick to bear repeating, but it worked every time. Parent birds sitting in caves rose and ambled forward like old folks in slippers shuffling out to get the Sunday paper. Snyder moved in behind the birds when the path to an egg was clear. Someone always followed him in with the incubator suitcase. Condors were expected to hiss and fly away. When they didn't, Snyder's colleague kept the parent birds away.

In 1983 and 1984, Snyder and the zombie patrol delivered a total of twelve condor eggs to the Los Angeles and San Diego zoos. Nine produced chicks; three never made it, though the failures weren't caused by handling.

"Nineteen eighty-four was perfect," Snyder said. "It was like, Oh my God, we've got peace in the world. We had five breeding pairs out in the wild, all of the funding we needed to do the monitoring, and the egg pickups were going beautifully." And it wasn't just Snyder who was feeling optimistic. In the fall of 1984, the Fish and Wildlife Service made the announcement everyone was waiting for: barring an unforeseen catastrophe, two or three captive condors would be released in the summer of 1985. If those releases worked out, more condors would be leaving the zoos in the summer of 1986. "We were set to go," Snyder said. "Everybody was optimistic, everybody was happy. We had no idea how short the moment would be."

━━━━━

Audubon promoted John Ogden out of the condor program late in 1984. After moving back to Florida, he started work as Audubon's new national director of science. Snyder and the members of the zombie patrol wished him well. Then they started getting ready to head back out into the field. When the egg-laying season began in January, they would look for the expected bumper crop to take back to the zoos, leaving the birds to double-clutch and pump out some replacements.

The first breeding pair was expected to produce an egg on January 15, 1985, based on the number of days that had passed since the birds were seen copulating. But when January 15 came, there was no sign of an egg or either of the birds. The next two eggs were due on the twentieth and the twenty-fifth of January, but on both days nothing happened. Snyder said it was at that point that he started to have a "terrible sense of impending doom." He and his colleagues had entered the Sespe expecting to find five eggs, but so far they had nothing. Frantic searches seemed to show that four adult condors disappeared over the winter, leaving just one breeding pair to keep the species going. Snyder thought it was the death knell for the wild bird.

"We needed to trap the free-flying birds and bring them into the zoos as quickly as possible," he said. "We had to assume that the missing birds were all now dead, and that the survivors were at risk. We had no idea what was going on out there in the wild. Leaving the birds alone would have been completely irresponsible."

Six condors died in 1985, including one from each of the four breeding pairs. That was 40 percent of the wild population. Only one of the missing birds was wearing a radio-tracking device, and it wasn't sending any signals. Most of the birds in captivity were now female; most of the birds left in the wild were now male. If you'd hired a group of expert marksmen and told them to doom the wild flock as efficiently as possible, these are the birds they would have hunted.

Snyder's colleagues on the egg team shared his sense of urgency. But at first Jesse Grantham, the biologist who'd taken Ogden's job, was not convinced a crisis was at hand. Back in New York, Grantham's former field boss seemed to hold a more suspicious version of that point of view. According to the minutes of a closed meeting at the headquarters of the Audubon Society, John Ogden and his colleagues still thought Snyder was just after power:

> The captive breeding people, strongly aided and abetted by Noel Snyder, are making this situation into a crisis and have requested that all wild condors be captured and put into zoos for "protective custody" . . . it is clear that this effort is designed to wrest control of the program away from the U.S. Fish and Wildlife Service and the National Audubon Society.

Ogden was the first to admit that the situation in the wild was extremely bleak. A single pair of condors produced a total of two eggs in the spring of 1985—nowhere near enough to trigger the expected release of a zoo-bred condor—and one of the eggs had been killed by a bacterial infection. Since 1982, at least seven condors had died in the wild, and now six more were gone. One of these birds had died after being shot *and* poisoned, and hopes for the others were dimming every day. The minutes of the closed meeting in New York showed that no one disagreed with that: "If we have lost three or four birds from the wild population, then the number of condors now in the wild must be only eleven or twelve."

Condor country had become a risky place for condors. Ogden knew that well. At the meeting in New York, Ogden said his radio teams had learned a lot about the habits of the wild condors, and a fair amount about what it was that might be killing them, but nowhere near enough to protect them. He saw an urgent need to learn as much as possible about the habits of the wild birds. The minutes of the meeting in New York seem to show that in the minds of the people who ran Audubon, this trumped the plan to beef up captive breeding.[6]

Starting immediately, increased emphasis in the field research program should be placed on very intensive surveillance of all wild condors on as near a day-to-day basis as
possible to try to discover the limiting factors acting on the
wild population, and for the development of a viable habitat reserve for the long-term survival of the wild population.
To this end we would suggest that all wild condors be
trapped and radioed in 1985.

Ogden hoped to keep the wild birds alive by feeding them at a
"bait station" on the Hopper Refuge. "Although relatively little is
still known about the present mortality factors, it is undeniable
that two condors that died in 1983 to 1984 died from something
they ate! If we do not get a handle on the problems facing the
wild flock, it would seem to us that we have probably lost the
species."

Ogden's plan was endorsed by the board of the National
Audubon Society and by the Washington office of the Fish and
Wildlife Service. Snyder's was endorsed by the California Department of Fish and Game, the Los Angeles and San Diego zoos, and
several prominent geneticists.

"That's when it really got ugly," said Jesse Grantham, who remained an ally of John Ogden's. "People started locking their doors
at the offices in Ventura about then. People on opposite sides of the
fight turned away when they passed in the hall."

Ogden and Snyder seemed to spend a lot of time on the phone
to Washington, with Ogden speaking mostly to Snyder's supervisors at a federal wildlife research center in Patuxent, Maryland, and
Snyder attempting to go over their heads by sending messages directly to the director of Fish and Wildlife. Snyder had been ordered
to stop holding public hearings, and staff morale had fallen

through the floor. Interim director Mike Scott brought in a team of psychiatrists. "My only rule in situations like those is 'thou shalt not sandbag,'" Scott said. "But when I joined the program I saw it happening all the time."

Snyder had a strong scientific case and no support from Washington. That case got much stronger when an extremely sick condor turned up in the foothills of the Sierra range. A delegation of veterinarians and ornithologists was sent out to investigate; the bird died shortly after they arrived. "One of the vets was holding it in his arms," Grantham said. "He told me he could see the light going out in the bird's eyes."

The political logjam broke in August 1995, when a dead condor was found on a ranch in southwest Kern County.

This condor had been half of the Santa Barbara breeding pair, which had just produced two healthy eggs in 1985. After it was found, the Washington office of the Fish and Wildlife Service approved an "emergency" plan to trap all of the remaining birds. Audubon did not give up the fight right away: the organization went to court to stop the trapping. Peter Berle, Audubon's new president, told reporters that the suit was filed because the trapping decision was "unilateral, was not supported by the administrative record, and was made by officials in Washington, who did not consult with their own scientific staff." A federal court agreed, blocking all forms of condor trapping.

More ugliness followed when a condor known as AC-3 was captured and tested for lead poisoning. When the tests came back, they showed that AC-3 had potentially lethal levels of lead in her blood. The court decision forced the recovery team to release AC-3 into the wild. She was sure to die if the ban remained in place. Audubon may have suspected a trick; they refused to back down. After that, Marsha Hobbs held a press conference at the zoo. She declared that

Audubon now had "blood on their hands."[7]

Peter Berle was appalled by these remarks. In a long, angry press release, he noted that Audubon had been fighting to save the condor for almost fifty years; he added that it was Audubon that originally called for the creation of a breeding program.

> Today's actions by Ms. Hobbs and the Zoo Association confirm reservations Audubon has had all along about the Greater Los Angeles Zoo Association—basically that it is more concerned with exploiting the birds and the recovery program to enhance its public and political standing than it is with saving condors . . . clearly, Audubon's longtime stake in this effort—scientific, financial and emotional—is too great to sit idly by and watch the condor become extinct.

Opposition to the trapping plan collapsed when the lawsuit was thrown out of court. That put an end to the injunction that had forced the trappers to unhand the birds.

twelve

ZOO

====

As soon as I saw the Vermilion Cliffs, I thought, "OK, we can stop looking. This is where we want to put them."

        —Mike Wallace, condor biologist, San
        Diego Zoo and Wild Animal Park

The condor in the kennel behind the front seat of the single-engine Cessna was alive and alert when the plane touched down at the small rural airport near the eastern edge of San Diego County. Don Sterner, a veterinarian at the San Diego Wild Animal Park, moved the kennel holding Igor to the front seat of his beat-up red Isuzu. Then he laid a blanket over the top of the kennel, hoping it would help keep Igor calm. During the windy drive to the park, Sterner kept raising the blanket slightly to peek into the kennel at Igor. By the time the Trooper reached the captive breeding compound behind the park, rumors of the taking of the last wild condor had begun to ricochet around the world. Reporters who'd been told that Igor was in San Diego were calling the zoo to ask for every last detail. Meanwhile, activists who'd threatened to "liberate" the bird were converging on the Los Angeles Zoo, having been led to believe that Igor was there.[1]

The guard waved Sterner through the big metal gates. After pulling up near the quarantine pen, he and a colleague slid the kennel out of the pickup, set it on the ground, and paused. *If something was going to go wrong it would happen here,* Sterner thought to himself. The door on this kennel was secured by eight metal latches that did not open easily, and when big birds rushed out too quickly, they'd been known to hurt themselves. A broken wing now would have the activists storming the gates of the compound. Condors also had a history of trying to make a break for it after scrambling out of kennels, doubling back into the faces of the keepers at the entrances to the pens. Lower on the list of concerns was the strong probability that Igor would try to bite off Sterner's fingers when he reached down to open the latches.

Another vet diverted the condor by pushing one of his boots up against the side of the plastic kennel. Almost instantly Igor's beak shot out between the metal bars, ripping into one of the leather-covered toes. Sterner got the latches open and started to pull on the door. Igor didn't let him to finish the job.

"It was like POW! and he was out of there," said Sterner. "He bounces into the quarantine pen, jumps up onto a perch, turns around to face us and freezes." The air sacs near his neck were all puffed out, and the skin on Igor's head and neck was bright, bright red. Every now and then the bird would jerk his head in the direction of a sound, only to end up looking at the plywood walls.

Igor couldn't see the other condors, but I'll bet he knew they were there. A guttural hiss from beyond the wall is all it would have taken, or maybe the stomping ruckus of a short mock charge. When the bird looked up he would have seen the sky on the other side of the thick black mesh stretched across the top of the pen. Clouds and tiny birds would have passed overhead, through his narrow field of vision. What an odd sight that must have been for him.

Igor was furious and terrified, and with good reason: in a matter of hours, his colossal range had shrunk to four hundred square feet. The pens that held the other breeding birds were bigger than that, but Igor was in quarantine, and the quarantine pen was twenty feet long by twenty feet wide. Giant wings and telescopic eyes were not going to do him any good in *here*.

On the other hand, Igor was alive, which is more than you could say for many of the condors he'd grown up with. "If we'd have left him out, he would have died," said Sterner. "There's no doubt about it. He would have eaten something full of lead pellets or pulled on a coyote trap and gotten blasted with cyanide, or been shot by somebody who didn't want him around. Or maybe it would be something we don't know about yet. We lost a lot of condors in the late eighties, and some of them just disappeared."

Sterner left Igor alone until the condor looked hungry. The bird's first meal was a processed loaf of horsemeat mixed with vitamins, bone, and odd bits of hair. Sterner then checked the bird's health, measuring body temperature, heart rate, and breaths per minute, and then picked through the feathers for lice. He soon looked up into the bird's nostrils and felt around inside the beak. Igor's blood would be tested for a huge range of abnormalities and toxins. If everything checked out okay, he'd be moved to a flight pen in thirty days.

"Condors are resilient birds," Sterner said. "They have to be, given what they eat. But they do get sick and listless, and we didn't want to let that in. We didn't need a virus running around in the captive flock, especially now. That could have been disastrous."

━━━━━

When Igor arrived at the San Diego Wild Animal Park on April 19, 1987, there were twenty-seven California condors left on Earth. Thirteen birds were living at the captive center in Los Angeles; the others were in San Diego. "We didn't have much of a population to work with," Sterner said. "We didn't know a lot about how the birds were related to each other. We were going to have to do some wild guessing when we put the first breeding pairs together. Basically, we'd be grabbing the females and putting them in with the males and hoping for the best."

California condor eggs had been hatched at these zoos when Igor was carried in the door, but none of the captive birds had ever tried to breed. Sterner and his colleagues were fairly sure it would happen eventually, but there were a lot of ifs. If the birds were all first cousins, the species was probably doomed; if the so-called crucial "founder birds" refused to breed, the species was definitely doomed. Sterner didn't know whether any of the birds were too old to breed or infertile. With only fourteen breeding pairs, he'd need a lot of luck.

At the Los Angeles Zoo, the situation was the same, except for the question of what to do with the condor known as Topa Topa. This was the bird Fred Sibley captured twice in the Sespe Sanctuary in the late 1960s, when it was refusing to eat and barely able to fly. At the Los Angeles Zoo, Topa Topa had been terrorizing keepers and other animals since the moment it arrived, and it now lived in isolation. "I don't think he knew he was a condor," said Mike Wallace, the man in charge of the L.A. breeding program. "He was always displaying to the keepers and they were encouraging it, because they thought it was special to be so close. If he wasn't in the mood to display, they stayed the hell away from him, because he was extremely aggressive. You could send a keeper in there with a net and Topa Topa would take it away from him. He was one angry bird." Wallace had made it a personal goal to see little Topa Topas brought into the world, but by 1987, all the other vulture experts thought he was wasting his time. The bird was old enough to feel the need to strut around in front of females, but when Wallace took his job in 1985, the females Topa Topa liked to strut for weren't condors, and some weren't even female.

In his years at the Los Angeles Zoo, Topa Topa had shared pen space with a lot of birds and animals, but never with another condor. When other California condors began arriving in the 1980s, Topa Topa was kept apart. "We didn't know what he would do," said Wallace. "And we couldn't risk losing a potential breeding bird." Wallace had a hunch that this was especially unfortunate from a genetic point of view, since Topa Topa had been taken out of the wild. He thought it possible that Topa Topa's genes were very different from the genes of the other captive condors, and in the long run, those differences could prove crucial. Topa Topa might turn out to be the only bird immune to a dreaded disease, or a condor that lacked a dormant but potentially deadly gene.

Wallace was encouraged to collect some of Topa Topa's sperm and try to use it to fertilize an egg inside another California condor. But vultures are notoriously unresponsive to that procedure, and Wallace didn't feel like giving up yet. "I told [the recovery team] I wanted to put a young Andean female condor in with Topa, and when they agreed, I went ahead and did it before anybody could change their mind."

As expected, Topa Topa flew right at the Andean, trying to intimidate it as he intimidated people. "He was going to fly right into her face and bite it," Wallace said, but the Andean stood her ground and watched Topa Topa come at her for a moment, seemingly frozen in place, and when the attacker was too close to change his course, she turned aside, so that Topa Topa sailed past her into a plywood corner. Wallace said the female then turned around and "beat the crap" out of her aggressor. "What I realized was that, *wow*, he knows a lot about how to manipulate people who come into his pen, but he knows nothing about his own species." Wallace left the birds together for a couple of days to see whether anything would change, but nothing did. Topa Topa kept attacking and the

Andean kept ducking out of the way, and when she needed to, she'd whack him with her beak, feet, and wings. "When he cornered her, she just thrashed him," Wallace said. "I don't think he ever touched her. After four days I was worried he would begin to get nervous about being around his own kind, so we pulled her out of there." Then he tried to make a different plan.

═══════

In 1987, the people who ran the nation's zoos weren't sure they wanted geneticists such as Oliver Ryder around. If they were, it was assumed that they would stay away from press conferences at which zoo directors bragged about the birth of the latest "pure" albino tiger cub, or the thirty-seventh chimp pumped out by the same set of adults. Those kinds of announcements may have thrilled the public, but they tended to infuriate Ryder, who tends to speak his mind.

"I'll give you an example," Ryder said when I met him in his office in the back of what had once been the morgue at the San Diego Zoo. "When AC-9 arrived at the Wild Animal Park in 1987, there was a big debate going on about the purpose of captive breeding programs. Zoo directors loved it when famous animals made babies, and that was all some of them wanted to know about. They didn't want to spend the extra time and money it was going to take to develop genetically useful breeding programs. They just wanted to talk about how cute the little critters were."

Ryder was one of the first to tell these zoo directors why they had to stop the nonsense. He was among the first "conservation geneticists" hired in the early 1980s, when the San Diego Zoological Society opened a research facility called the Center for Reproduction of Endangered Species, or CRES. When I first met Ryder in the late 1980s, CRES had already established itself as a world-class center for animal research. Projects under way in the former morgue ranged from the creation of an aluminum water bed for pregnant mammals, to the installation of hormone pumps in animals with breeding problems. Scientific papers reported on the pathology, virology, endocrinology, behavior, and genetic diversity of a wide range of very rare animals.

In the late eighties, molecular geneticists weren't always media-friendly. But Ryder was, and is. Television crews were always dropping in at CRES to film the fog of liquid nitrogen that poured from the top of what Ryder called the "frozen zoo"—three metal tanks full of semen, egg, and tissue samples taken from endangered animals and stored in a supercharged freezer.

"Three hundred fifty species," he said as we watched the fog. "Three hundred eighty-five degrees below zero. Fahrenheit."

"What got you into this line of work?" I later asked, walking back to his office. Earlier that day, one of the other conservation geneticists at CRES had cracked a bitter-sounding joke about how rich he'd be if he'd taken his mother's advice and studied the human genome.

"I was a stamp collector when I was a kid, and when the postal service put out a condor stamp, I thought it was very cool. After that my parents used to drive me up to Griffith Park to see the condor at the L.A. Zoo. The zoo wasn't much to look at in those days—I don't remember much besides the cages and the carousel—but there was this one cage that had a condor in it, and it was Topa Topa."

Ryder interrupted himself to dig around inside a small refrigerator. Eventually he pulled out a tiny glass vial and handed it to me. "Look," he said. "California condor DNA, in a gel. I've got the DNA of every living condor in the world in this refrigerator, and from many of the dead ones. When we look at that gel, we see patterns of bands that tell us all kinds of things."

I gave the vial a studious look, lamely attempting to hide the fact that I had absolutely no idea what he was talking about. Then I handed the vial back to Dr. Ryder, who returned it to the refrigerator. This was what the condor was in 1987—bits of DNA in vials lined up in a fridge that belonged in a college dorm room.

"The lines look a little like bar codes," he explained. "When you compare them you find out how related the individuals are. It's a complicated process that involves the use of mathematical algorithms that in my warmest moments I can only come close to understanding, but the basic idea behind the math is that you want to come in and count the band-sharing while giving extra attention to the anomalous bands. When you find a bunch of bar codes with the same irregularities, it's likely that you're looking at condors with a common ancestor, a shared line of descent."

Over the course of the next few years, this kind of work would become routine, and not just at the zoos. Lawyers would be throwing around phrases like "shared anomalies" in bitter paternity disputes; eventually historians would join the fray. According to Ryder, it was the presence of an odd snip at the end of a long string of Y-chromosomes that proved that Thomas Jefferson made babies with one of his slaves. But that all lay ahead in the 1980s, when Ryder and a colleague set out to measure the genetic relatedness of the last of the California condors. Back then, nobody had ever done this kind of work before. As a result, fights over whether zoos should breed for numbers or genetic diversity didn't make much difference in the end. Without a family tree to refer to, zookeepers could only guess at whether breeding pairs were totally unrelated, barely related, closely related, or ridiculously inbred.

Ryder and a colleague made those arguments matter when they found a way to draw the missing family trees. The condor project needed this breakthrough; when it was done, the future of the species seemed a little brighter than it had before. According to the bar codes, the last surviving condors hadn't bred much with cousins, siblings, parents, or close relatives, and this raised the odds that the condor genome was in pretty good shape. It probably wasn't full of recessive genes that worked like hidden time bombs set to blow up somewhere down the line, triggering killer diseases and severe deformities.

"The birds fell into three distinct genetic clans," said Ryder. "This was a good thing. One of the clans held the condors we already knew were related, and this was also good. Topa Topa and AC-9 were in one of the other clans. They were from the same family, which was cool."

"So, it's possible to argue that this was a turning point in the history of conservation biology for biologists, right?"

"I'm fond of saying there's a dividing line in the history of biology," he said, ducking the question, "with the pregenomics era on one side and the postgenomics era on the other. In 1987, we were stepping across the line, into a world where everything was going to be seen through the lens of genome biology. And that's exactly what happened. Everywhere you look, there are new tools, new technologies, new and wonderful opportunities. The problem is that when you look at the objects of your concern—species diversity, deeper gene pools, that kind of thing—all you see are trend lines pointing down toward extinction. We want to use the encouraging curve to affect the depressing one."

The condor work proved it could be done.

Meanwhile, back at the sanctuary, Lloyd Kiff and Robert Mesta were revising the Condor Recovery plan and guarding the habitat. Twenty-two days after Igor's capture, the Forest Service decided to review the status of 383,867 acres of lands withdrawn from leasing and mining operations to protect the California condor, raising the possibility that much of this land, including the Sespe Condor Sanctuary, would be reopened to mining companies, timber companies, home builders, road builders, motocross riders, and others. Kiff and Mesta were doing what they could to keep the lands withdrawn, but George Bush Senior's White House was extremely sympathetic to the needs of the extractive industries, and the state of California didn't seem interested in getting in the president's way. In 1986, a spokesman for the Department of the Interior had wondered aloud about whether the condor program had become a biological "lost cause" and a financial "black hole." The overall bill for the condor program had just reached $20 million, and a lot of people thought that was too much.

One of the loudest critics of the program was Bil Gilbert, a well-known nature writer who'd been following the effort to save the bird for many years. In an article he wrote in 1986 for *Discover* magazine, Gilbert announced that he'd had enough of the endlessly melodramatic fight to save not one but "two of the most insignificant of animals" from almost certain extinction. The animals in question, Gilbert wrote, were the black-footed ferret and the condor. "The condor has become a kind of Charles and Di story," he added, "what with breathless bulletins about little buster's hatching at the San Diego Zoo, that adjunct of the Johnny Carson show."

The point of the article was in the headline itself: WHY DON'T WE PULL THE PLUG ON THE CONDOR AND FERRET? In the piece, Gilbert argued that "in the strictly zoological context" it was hard to justify spending a cent on "gaudy crisis patients" that had nothing to do with the natural world around them. Gilbert seemed to think that this was a fate far worse than mere extinction; "managed" species bred in captivity weren't worth saving, he implied. "Even if all goes very well, we will, in fact, have produced these animals as we produce poodles, and we will manage them like domestic stock."[2]

So why not spend the next $20 million on ecologically important mountaintops, or prize wetlands full of rare animals and plants? Why waste all that money on losers like the ferret and the condor? Gilbert planned to write something mournful when the ferret and the vulture were gone, even though that "might not be such a terrible thing as is commonly, reflexively thought":

As has been noted, the functional importance of the condor and the ferret is now symbolic, ecological. The passing of two such celebrated species would dramatically call attention to the process of extermination, and the ways our activities now influence the process. Also if the condor and the ferret were to go we might be more greatly motivated on behalf of other species that are headed in that direction.

As the leader of a team of scientists created to advise the condor field crews, Lloyd Kiff was always sending letters to the editors of magazines that printed stories about the condor. After reading Bil Gilbert's "Pull the Plug" article, he felt like firing off a long one, in which he would have pointed out the following:

1. The condor program spends a lot of money buying property important to the future of the species.

2. Politically, there's no chance that money taken from the effort to save condors would be used to save lesser-known species.

3. The condor population in the zoos is getting bigger as more eggs are laid; someday soon, if all goes well, birds hatched and reared in captivity will be released in Southern California.

4. When you save a big chunk of habitat for the California condor, you save it for hundreds of smaller species at the same
   time.

But Kiff was way too busy to write letters like those, at least until
he finished revising the official condor recovery plan. A plan can't
be accepted as official until everyone involved in the process signs
off on it, and then waits for everyone else to sign before reading the
plan again. If the person busy rereading the plan finds important
language that wasn't in the plan the first time around, he or she can
take the offensive language out and start the process over.

Kiff had signed off on the recovery plan after taking out a paragraph that would have made it easier for the owners of a former
land grant on the far side of the mountains to build suburban
homes in an important corner of the condor's feeding range. After
Igor's capture, lawyers representing the Tejon Ranch began to argue
that the condor had become "technically extinct" when the last bird
left the wild; the lawyers also argued that because the Endangered
Species Act did not protect the habitats of extinct species, then it
couldn't be used to protect the habitats of "technically extinct"
creatures.[3]

These claims had never been tested in court, but Kiff had a feeling they were related to an onerous paragraph he kept removing
from the new version of the condor recovery plan, only to see that
it had been restored when he read the plan for what was supposed
to be the final time.

The paragraph would have exempted the owners of the sprawl-
ing Tejon Ranch from all land-use rules designed to protect the
condor and its range. "Every time I found it I'd cross it out and
send it back" to the Ventura office of the U.S. Fish and Wildlife Service, he said. "They would send it off to Washington for a final re-

view, and when it returned, the paragraph would be back in it. I don't how many times I killed it before Washington finally gave up," Kiff said. "It was like trying to stomp on a cockroach."

In 1989, these gathering land-use fights were postponed by a controversial experiment. Over the objections of a number of ornithologists and activists, female Andean condors raised in zoos were released in the Sespe Condor Sanctuary. Lloyd Kiff called it a "practice run" designed to help the field crews prepare themselves for the day when the Andeans would be replaced by zoo-bred California condors. The critics called it "foolish and counterproductive," warning that it might end up delaying the return of the California birds, and adding that Andean condors, unlike their distant cousins, occasionally ate from the carcasses of animals they'd killed themselves. Rumors that the Andeans would never be removed began to circulate, along with talk that the California birds would be moved to the Grand Canyon.[4]

Seven Andean nestlings were released in 1989; six more followed in 1990. Five were captured and then rereleased after badly fastened radio transmitters tore up the edges of their wings; one was removed because it never quite learned to fly; and one was electrocuted when it crashed into a power line.

Then, with one exception, the other birds began to see the wind, staying out of trouble while gradually expanding their range. The last of the Andeans to fail had a thing for hang gliders. More than once, shocked pilots turned their heads to see a vulture with a ten-foot wingspan pull up next to them, touch the wing of the glider with the tip of a primary feather, and fly at the same exact speed for a couple of minutes, all the while staring into the eyes of the human with the nylon wings. By the time the experiment was over, the remaining Andeans were flying through the mountains in eastern Santa Barbara County on a fairly regular basis, and one traveled

ninety miles east to Riverside County. The field biologists were
pleased to see the Andeans eat almost exclusively from "clean" carcasses laid out for them; they weren't so pleased when big black
bears stole the carcasses in the night.

I saw the Andeans in the Sespe in 1990. Dave Clendenen, then
the federal government's senior condor biologist in the sanctuary,
led me to a plywood observation blind near the place called Koford's O.P. I was looking out across a dry wash when I started hearing beeping noises coming out of the tracking gear. Clendenen
tapped me on the shoulder and pointed up at the opening in the
top of the blind, where two Andeans floated a hundred feet above
our heads.

"You've been busted," said Clendenen, stating the obvious.
Shortly afterward, the condors swirled up and over to the south, toward the smog.

———

Early April, 1988: Don Sterner and the other keepers sat in the captive breeding center, staring at the closed-circuit monitor connected to the camera in UN-1's pen. The condor was sitting in the
corner of the pen doing nothing of significance, but all eyes remained on the screen. UN-1 was the first of the female Californians
to copulate that year, and lately she'd been moving lugubriously
around the pen. If this bird was pregnant, this was probably her
due date, since it was about five months since she had mated with a
condor known as AC-4. For that reason, every time it looked as if
UN-1 was about to move, all the keepers watching the monitors
leaned forward and held their breath.[5]

"We're all wondering whether she's sitting on an egg," Sterner
said, "but none of us can see one. We're waiting and waiting in
front of these monitors and no one is saying a word, and the atmo-

sphere in there was so incredibly tense I couldn't believe it, and then she moved."

UN-1 stood halfway up and sat back down again, readjusting her weight. Everybody saw the egg flash white beneath her. For a moment, the walls of the trailer shook with the cheering of the keepers; then AC-4 climbed up into the nest box. "I said, 'Okay, everybody, let's go get that egg,' and we rushed over to the pen," Sterner added. "We netted the birds and put the egg in a bucket of warm finch seed and delicately carried it back to the incubation chambers and then we took one of those green kitchen sponge pads and sanded off the stuff that was stuck to the outer part of the shell and looked for cracks."

No cracks. Sterner held the ten-ounce condor egg over a bright light called a candler and stared at the yellow-orange glow. "Everything inside looked fine," he said. Seconds later, he put the egg down on a padded shelf in the middle of a boxy wooden incubator—98 degrees, adjustable humidity, backup power supply—and called Art Risser, the zoo's curator of birds. More jumping and cheering.

Sterner and the other keepers checked the egg every day after that, rocking it back and forth on its longitudinal axis to distribute the heat from the candle, tracing the growth of the air sac with a soft lead pencil that was never pointed straight at the shell, adjusting the humidity of the incubator to speed up or slow down the rate at which water in the egg was evaporating through the shell, and looking for the bump on the surface of the yolk that is the first sign of cell division. When Sterner saw the bump, he picked up his walkie-talkie and made a short announcement to the staff at the park: "ATTENTION ANIMAL SCIENCES UNITS, THE CONDOR EGG IS FERTILE." People called him back to yell and scream.

This was the first fertile California condor egg ever laid in a captive setting. Brief accounts of the grand event appeared in papers around the world. Tape recordings of vulture grunts and hissing sounds were played nonstop in the incubation room for the next several weeks. On April 26, a tiny hole called a "pip mark" appeared in the side of the egg. More stories ran as the chick started hatching. Then the chick got stuck.

"We dressed up in our surgical gear and went back in with nervous hands," Sterner said, "knowing that the world would be out there waiting while we broke this little guy out."

This is called "assisted hatching," and it doesn't always work. Sterner and biologist Cindy Kuehler took turns chipping at tiny fragments of eggshell, trying hard to stay away from parts of the membrane that could rupture and kill the chick. When they were sure the membrane was safe, they broke off slightly bigger bits of egg with their gloved fingers and a set of tweezers. "Steady gentle pressure is the key," said Sterner. "Condor shells are kind of thick."

Sixty-one hours after the pip appeared, at 5:38 P.M. on April 29, 1988, Sterner poured a 6.5-ounce condor chick named Molloko out of a meticulously disassembled egg into Kuehler's hands. "The little legs were really kicking hard," he said. "When we were done we took a group photo and went to a big press conference. Then we drank a lot of wine."

Molloko was the only condor chick produced that year, but it was all the zoos needed. Critics fell silent and reporters fawned. Four California condor chicks hatched in 1989; eight more arrived in 1990. Keepers took the eggs from the parent birds as soon as they were laid and raised the chicks with puppets meant to imitate adult condors and fool the offspring. Many of the breeding pairs double-clutched, producing second eggs, which were also taken away. Some of the breeding pairs replaced the replacement.

By the end of 1990, there were forty condors living at the zoos, thirteen more than there were in 1987. By the summer of 1991, the count was up to fifty-two. By the standards of a species that was used to producing one egg every other year, this was miraculously rapid growth. The Andean condors that had been released in the Sespe had all been trapped and flown to South America, where they were released for the last time. The glitches in this process got no attention, which may have been a big mistake. Misbehaving Andeans were dismissed as freakish; a deadly collision with a power line was considered not likely to happen again, when in fact it should have been. Finally, at the San Diego Wild Animal Park, keeper Bill Toone said the condor puppets used to feed and preen the chicks weren't fooling anybody. "It only took the chicks a few days to figure out that there were people behind the puppets," he said. "One minute they're oblivious, and then they're looking at you through the one-way mirror."

But Mike Wallace at the Los Angeles Zoo didn't share Toone's fears, and by 1990, he'd become the man in charge of the reintroduction program. Wallace felt an urgent need to put the birds back where they'd come from, and he hadn't noticed any serious problems at his end of the breeding program. At the Los Angeles Zoo, even Topa Topa was trying to get into the act. Wallace said he saw his former problem bird start struggling with his sexuality in the summer of 1989, when it did a mating dance for the bushes in its flight pen (later it attempted to mount those same bushes). Topa Topa also tried to incubate a stray piece of PVC pipe that year. Wallace considered these unusual events an improvement over Topa Topa's previous attempts to mate with the keepers, so he put a new female condor in the flight pen. Topa Topa preferred the bushes at first, but gradually he came around. In 1990, he attempted to breed with the other condor in his pen. When the egg turned out to be infertile, Wallace switched it with the egg of an Andean condor. When that egg hatched, Topa Topa was acting like a regular old proud papa, feeding the chick in twenty minutes.

"We let him rear the chick for about four months before we took it away," Wallace said. "And he was a model dad. He would lie there and the thing would crawl on him, and he was gentle, great. That winter he mounted a female condor named Malibu and did everything he was supposed to do perfectly. The egg was viable, the chick hatched, and the parents have been happy and fertile ever since."

———

On January 27, 1992, two California condors were released in the mountains of south-central California. Each had hatched from an egg produced by Igor and the condor known as the Matriarch, or AC-8. When the birds emerged from a man-made nest on a rocky promontory in the Los Padres National Forest, naturalists standing on distant cliffs applauded and drank champagne. The birds started jumping up and down and flapping their giant wings. It looked like they were dancing.

Pictures of the zoo-bred California condors leaving the release pen appeared all over the country. The next day, virtually all the coverage was positive. The zoos felt a sense of vindication.

Lloyd Kiff, the leader of the Condor Advisory Team, reacted differently. He'd signed the documents that ordered Pete Bloom to capture the last of the wild condors back in 1987, and at some level, he'd been worrying ever since. The worries weren't particular, but they'd nagged at him for years. "I didn't want to be remembered for signing those papers," he explained. "When I saw the birds take flight I thought, 'Okay, I'm off the hook.' I was thrilled—we all were—but what I felt most was relief."

More than a hundred California condors have been released to the wild since that day in 1992. But it turns out the first two California birds released didn't last very long. One was captured and returned to a zoo when it started acting tame. The other died after slurping up a puddleful of bright green antifreeze.

thirteen

GRAND CANYON

═══

Cage-raised birds will have to learn the vast geography of their habitat, techniques of flight, actions to avoid injury by storms or becalming by wind failure, methods of finding food and coping with Golden eagles. . . . Few will survive to the breeding age of at least six years.

> —Carl Koford, Museum of Vertebrate
> Zoology, University of California,
> Berkeley, February 1979

When Robert Mesta saw himself hanging in effigy from the tree near the high-school auditorium in Kanab, Utah, he took it as a bad sign. Then he entered the auditorium and saw the stack of weapons near the door. Deputies in riot gear were asking grim-faced locals to set aside their knives and pistols before stepping through the metal detector and into the hearing room. The weapons would be returned to them when the yelling and screaming were over.[1]

Mesta is a soft-spoken Native American raptor biologist. In 1996, he was running the condor reintroduction program, but he hadn't met many of the people waiting in this auditorium. Most of them were hardworking nondrinking Mormons who would never even think of firing a gun at a public hearing in a high school, but somebody had called a threat in to the sheriff, and he wasn't taking any chances.

"It was not what I expected," said Mesta. "The proposal had been out for a while by then, and we hadn't heard any serious complaints."

"The proposal" was a plan to return the condor to the Grand Canyon, or someplace near the boundaries of the national park. The hope was that the birds would eventually return to the caves and canyons they'd left ten thousand years ago, after the Pleistocene blitzkrieg.

Condor experts called the reintroduction plan vital for many reasons, starting with the relatively urgent need to buy some catastrophe insurance. Horrible diseases and other freak events are always a threat to isolated groups of critically endangered species, and zoo-bred condors released in California fit that profile. By building up a second free-flying population, the condor keepers would reduce the chance of a "kill-them-all-in-one-fell-swoop" disaster event to 0.000 percent, or to a number so close to that, it wasn't worth calculating. Simultaneous regional disasters might be enough to erase both groups of wild birds at once, but how likely could that be? When was the last time anything happened in two fell swoops?

If you needed arguments of that sort, Mesta was your man. In California, he was known as an official who listened when other people talked, and for his ability to keep small fights from turning into big ones. Here in Arizona, he was ready to argue the case for reintroduction on any number of grounds, ranging from a surge in tourism to the need to keep the condor's genes diverse.

When it helped, Mesta even pulled a kind of cosmic shame card out of his uniformed sleeve, arguing that condors might still be in the canyons if it weren't for us human hunters, and now here we are ten thousand years later with a chance to make things right. As crazy as it sounded, the people of southern Utah and northern Arizona were now in a position to make the Grand Canyon look a little bit grander. When was the last time that had happened? Putting the condor back where it belonged was the right thing to do, wasn't it?

Of course it was.

Some of Mesta's colleagues thought the plan was as good as approved. They'd been flying in from California for years, quietly examining a series of release sites before agreeing to set the birds free on the uppermost lip of a 1,200-foot-tall curtain of rock called the Vermilion Cliffs. This place was in Vermilion Cliffs National Monument, adjacent to the national park, but that didn't matter to the experts. Mike Wallace remembered gazing down on these cliffs from the window of a plane and thinking, *We can stop looking now, this is perfect, this is the place we want to be.*

Hearings in Phoenix and Salt Lake City had been uneventful. Mesta let his guard down after that, which was a big mistake. Mormon kids everywhere—me included—have always been raised on stories of pioneering martyrdom and federal oppression, but in southern Utah and northern Arizona the wariness can be tectonic. By some accounts, this wariness has been a local trait since the late 1800s. That's when the residents of Kanab helped a fugitive Mormon named John D. Lee avoid federal posses, even though the leaders of the Mormon Church had identified Lee as the leader of a group of Mormons that had stopped a passing wagon train and killed almost everyone in it in September 1857. Lee denied the charge and claimed that he'd been made a scapegoat for what's now known as "The Mountain Meadows Massacre." By some accounts, that Mormon sense of "us against the world" has been periodically reinforced ever since, most notably in the eighties, when the federal government began restricting grazing on badly damaged federal lands. Not long afterward, David Brower sank a plan to transform the region forever by building four new dams on the Colorado River.

The uranium mine in the town of Kanab went out of business in the seventies, to be followed shortly after that by the local timber mill. Global competition was the problem, but you wouldn't hear that from the newly unemployed. They blamed meddlesome federal bureaucrats who never kept their word.

They also blamed the Endangered Species Act. Elected officials fumed when logging was restricted near trees deemed essential to the long-term survival of the Mexican spotted owl, and when reintroduced gray wolves ran off with the occasional calf. Environmental groups that local folks had never heard of before wanted more restrictions, and at one point there was talk that the forests would be closed to protect the northern goshawk. Every time the residents of northern Arizona turned around, there seemed to be another rare critter on the land, making it harder to dig the mines and pay those grazing fees.

This was the mood in Fredonia when the condor plan was officially unveiled in 1995. Joy Jordan, mayor of Fredonia, remembers sitting down to read the version of the plan that was published in the *Federal Register*. Jordan says she was shocked by the paragraph describing the range of this protected bird. "We were all inside the habitat lines," she said. "That was a fearful thing. No matter where we went or what we did, the condor would outrank us."[2]

That wasn't true, but it seemed like it. It was also scary. To Jordan and her allies in Fredonia and Kanab, the condor wasn't so much a bird as it was a giant floating regulatory mechanism. Federal bureaucrats could use it to justify pretty much anything they wanted, she thought. That was not acceptable to Jordan, or to her friends and neighbors.

The result was a hearing that very nearly killed the reintroduction project. It started when Mesta, the federal biologist, stepped up onto the small wooden stage at the Kanab High School auditorium and sat down in a chair that faced the audience. Sitting next to Mesta were Mike Wallace of the Los Angeles Zoo and Bill Heinrich of the Peregrine Fund, the nonprofit group that had been hired to

manage the condor restoration program. At the center of the stage
was a federal "facilitator" brought in to keep the hearing moving.
He didn't know a thing about the condor or the plan.

The meeting was ugly from the start. Ranchers, miners, loggers,
and others in the audience lined up behind the microphone, taking
turns denouncing the condor plan in the strongest possible terms.
They'd heard all about the economic damage condors were capable
of doing, and they didn't think the birds were worth the trouble.
They knew you could go to jail for shooting a condor, even if it was
an accident, even if one should happen to swoop out of nowhere
and hit the grill of a pickup truck. They did not want nosy federal
officials chasing condors back and forth across the grazing lands,
nor did they look forward to the patronizing lectures they would
get from power-mad field biologists half their age.

Mesta badly wanted to respond to some of this, but that was
against the ridiculous rules that govern these kinds of hearings. The
only official allowed to talk was the hired moderator, and he didn't
know enough about the birds to help.

Mesta could see that no one in the audience understood these
rules—after airing grievances, they waited for him to start talking.
When he didn't, they got even madder and asked more questions to
which he could not respond. In the end, a few public hints were
dropped to say that if the condors were released they would end up
full of bullets.

Some of the speakers had been waiting for decades for the
chance to tell the government off. Mesta listened while the locals
complained about the problems they were having coping with
other endangered species, the grudges they held against other federal officials, and a lot of unrelated things. One local man complained about the damage done by nuclear-weapons tests in the
Utah desert in the fifties.

"They were holding us accountable for everything bad that had ever happened to them," Mesta said. "They felt like their complaints had been ignored for years, and I think they had a point." Mesta said three people spoke in favor of the condor plan. One was a high-school science teacher, who said he felt like he'd just walked into the Roman Coliseum.

Shortly after the meeting in Kanab, the Fish and Wildlife Service put the reintroduction plan on indefinite hold. For the next several months, Mesta and his colleagues sat down with everyone who felt like sitting down, talking, and then listening. After the formal public meetings ended, they followed the lines of pickup trucks to local restaurants, and went inside to listen more. Over and over, local folks told them it wasn't a personal thing, or even a condor thing. They just didn't want the feds around.

Mesta could feel them turning. What he couldn't feel was whether they were turning fast enough. U.S. senators and representatives from Utah and Arizona were putting pressure on Interior Secretary Bruce Babbitt, requesting further studies, additional limitations, stronger reassurances. It didn't help that Arizona was then governed by a Republican hard-liner named Fife Symington, who once implied that the best way to comfort an endangered species was to shoot it.

Secretary Babbitt's advisers were eager to throw some water on this fire, lest it spread to other programs. They were also forced to deal with some tricky legal questions. For instance, if a bird leaves its home for a millennia, does the place it left still qualify as home? If the answer was yes, was there *any* place the government couldn't put endangered species?

A self-described "good bureaucrat" named Bruce Palmer helped the government jump that trip wire. Even though he ran the Southwest office of the Fish and Wildlife Service, Palmer wasn't brought into the condor debate until after the meeting in Kanab. "Basically I got a call that said 'Get out there and put out the fire,'" he said.

With the help of Robert Mesta, he moved in and quietly won the arguments against reintroduction. When a spokesman for the pilots who flew tourists over the canyon complained about possible collisions, Palmer told him it was a good point and "now all you have to do is prove that you've hit birds over the canyon before." The pilots of course knew that if they documented that such collisions had happened, they might have some canceled flights; shortly afterward, they dropped their challenge to the reintroduction plan.

Another objection fell to a more cunning bureaucratic maneuver. Rather than get involved in a fight over whether the birds had been away long enough to lose their residency status, Palmer simply noted that according to the laws of the state of Arizona, plants and animals included on the state list of endangered species were automatically eligible for federal protection, and the California condor was definitely on the state's list. Palmer knew that this was true because he'd personally added condors to the list a few years earlier, before the plan to put the birds in the Grand Canyon had gone public. No one had objected at the time and now it was the law.

And so it came to pass that in the fall of 1996, a twin-engine plane that was normally used to fight forest fires took off from the Burbank, California, airport. Inside the plane sat six condors inside six kennels tightly strapped into the cargo bay—one of the pilots later said it looked like the birds were preparing to parachute out of the plane above the canyon.

When the plane landed in Arizona in the vicinity of the Vermilion Cliffs, Mesta and some other state officials moved the kennels to a helicopter, which then flew to the top of the cliffs. From there the men moved the kennels to the back of a truck, which took the birds up a gully of a road that led to the top of the cliffs where there was a large release pen.

At the bottom of the cliffs, the Bureau of Land Management had regraded the road and built a podium. To accommodate the gathering crowd, officials had stretched a line of Porta Potties behind the podium.

Seven hundred people came to see the birds fly. Forunately, the release went off without a hitch. Interior Secretary Babbitt hailed the grandeur of it all. Republican Senator John McCain said a few words of his own. Arizona's Governor Symington, soon to be impeached on unrelated charges and then forced to resign, said he thought there might be a place for the condor in Arizona after all. Joy Jordan gave Bruce Palmer a hug and told him the birds were beautiful. After a few hours everyone went home and the podium was dismantled.

———

Winter 1999: Eleven hundred feet above the spot where the condors were released, Shawn Farry parked a Peregrine Fund pickup truck in a clearing near the edge of the Vermilion Cliffs. When it's not snowing up there, the rocky red landscape looks unforgiving and otherworldly, like the surface of Mars, only with trees. But on this particular day it was snowing hard, and that made the landscape look lethal. One false step on the slippery rocks and you could easily be lying on your back 1,100 feet below.

But Farry had the pathway wired. He not only walked it on a regular basis while tracking the tagged condors, he often walked in the dark with a hundred-pound calf carcass on his back from the truck to the edge of the cliffs, where he bolted the carcass to the slab called Prometheus' Rock. Several condors slept on ledges just below that rock, tucking their heads in under their wings for warmth and protection. In the morning, the leathery heads would sprout back out of the black-feathered bodies, and then the birds would spread their wings and rise above the edge of the cliff like so many flying saucers. Farry would be watching from his spot inside the torture chamber that was the blind.

"The blind isn't insulated, and when the little window in the front is open, the wind and snow blast right through and smash you in the face," he recalled. "I'm sitting in my sleeping bag, wearing four or five layers of clothing. I've got a small propane heater—it doesn't do any good unless I take my feet out of the sleeping bag. The birds are a hundred meters ahead of me, out on the edge of the cliff, but the snow is heavy and I'm having trouble seeing them. Their legs look way too gray to me, and I'm worried they might have frostbite, but I could be wrong. I put my head on the bottom of the blind so I can look through some cracks in the wood and see the feet a little better. Maybe the legs are only white from condor shit."

From 1997 until the start of 2002, Farry was the lead biologist for the Arizona reintroduction program. He'd never worked with condors before, but he'd read a lot about them when he was a kid growing up in upstate New York. He said he was depressed for weeks when Pete Bloom caught the last of the free-flying condors in California in 1987. He said he was intimidated the first time he saw a condor, at the San Diego Wild Animal Park.

"It was like you didn't want to look them directly in the eye," said Farry. "Like you were supposed to look at the ground instead. The next thing I know I'm sitting in this blind on this cliff in a snowstorm, with the fate of the Arizona program in my hands."

Local ranchers thought he was crazy. But gradually he'd won them over. They had watched him drag the carcasses of stillborn calves for miles over the rocks, just in case the condors needed something fresh to eat. The ranchers had invited him into their homes, eating Sunday meals. They knew he was going to keep these birds alive or die trying, and that had earned him some respect.

But Farry and the condors still had enemies. He and his colleagues still changed the subject when strangers asked them what they did for a living. They still had to stand there and take it every now and then when a gas station attendant or a bartender launched into a tirade against endangered species.

They also got some nasty hate mail. Farry read some of it while he was sitting in the blind, waiting for the storms to clear. "Meet your worst nightmare," one missive said.

You guys are a bunch of criminals! You are worse than the vultures you feel give meaning to your pathetic lives! I will not rest until every damned condor is removed from the Grand Canyon! I will not sleep until every newspaper in the country knows the truth!!! [I am] fed up with you, you environmental wackos!

That was the worst of a long series of e-mails written by a man who had helped an injured friend hike out of a campsite at the bottom of the Grand Canyon. While he and the friend were making their way up and out of the canyon, a group of condors had found the site and ripped it to pieces. Farry's crew found the wreckage, cleaned up the campsite, and started packing the detritus out of the canyon. On the trail, they met the man who had just helped his injured friend hike out. Farry said the guy went ballistic.

"He wanted to be covered for the damage," Farry said. "But the Fish and Wildlife Service couldn't do that. Campers in the canyon are supposed to know that they can't leave their gear out in the open, and if they have to do it, they are taking their chances. If we paid this guy, we'd end up paying all kinds of people, and that wasn't going to work. Do you pay a guy who gets bit by an ant and goes into shock? Do you pay a guy when a passing bird craps on his car? It would never end."

New groups of condors were released on a regular basis, and soon there were dozens of condors soaring over the area. With the help of five colleagues, Farry worked insanely hard to keep the birds alive. Each bird had a pair of transmitters on its wings, and Farry and his coworkers were expected to follow the birds' movements twenty-four hours a day. When a condor ate something that hadn't been laid out for it, Farry was supposed to backtrack and find out what the carcass was. When a condor buzzed a crowd on the South Rim of the Grand Canyon, Farry was supposed to catch the bird, or chase it away. He wore out a series of pickup trucks chasing the six birds around. When a bird was injured, he tracked it down, and put it in a kennel, and drove the kennel to the office of a friendly veterinarian in Page, Arizona.[3]

"You want surreal? Here's surreal," Farry said. "I'm sitting in the waiting room of the animal hospital in Page, reading a magazine and waiting for a space in front of the X-ray machine, and the kennel starts hissing and bouncing toward the door. A giant beak would poke out through the bars. People would start freaking out. Usually I just kept reading."

Farry said it wasn't long before he figured out that it was physically impossible to manage these birds and study them at the same time. It took hours to drive from the Vermilion Cliffs to the South Rim. Also, if a condor got especially sick, you had to drive it all the way to Phoenix, which sometimes meant driving for seven hours straight with a condor vomiting out of the sides of the kennel next to you.

Late in 1999, one of Farry's condors was shot to death and left on the ground where it had fallen. His first thought was that an enemy of the birds had finally acted on a threat. Then he got a call from a man who had once been one of the most vocal enemies of the condor program, who said his fellow ranchers and some miners wanted to put up a reward for information leading to the capture of the condor killer. The shooter turned himself in a few days later, claiming he'd shot the bird by mistake. He paid a small fine and did some fieldwork.

"We didn't have very many enemies in the end," Farry said. "And to be honest, the real yahoos were not the local ranchers and residents. The local folks we dealt with were good, hardworking, honest individuals who initially opposed us for a variety of reasons, and then changed their minds."

Farry thought the phone call marked a turning point for the reintroduction program. After that, when strangers asked him what he did for a living, he told them.

fourteen

NOT THE SAME BIRD

Condor country is not safe for condors anymore.
— Noel Snyder, condor biologist, 2002

When condors bred in zoos don't act like condors in the wild, Les Reid tends to gloat. "We told the zoos not to lock the condors up," he said. "We said, 'Hey, leave them out in the wild so they can teach the young condors to survive.' They said 'Oh no, we can't do that. We've spent God knows how many millions of dollars setting up all of these captive propagation facilities where we're raising young birds with puppets and such, and we don't want to waste that money now, do we?'"

Reid is a retired pipe fitter and a former member of the board of directors of the Sierra Club. With his wife, Sally, he helped ensure the protection of huge parts of the condor's range in the early 1960s by goading the California state legislature into passing a historic wilderness protection law. Reid said he did it partly to save the condors he and his wife often saw while hiking in the mountains of south-central California. "Those were the real birds," Reid said. "Not like now. Two miles up in the sky, barely moving, none of those radio gizmos hanging off their wings. That's the way you're supposed to see a condor."

The Reids used to argue that condors had certain rights. First among them was the right to lead wild and unfettered lives. In retrospect, it's more than likely that this approach would have led directly to extinction, but Reid is not the kind of guy who tends to give a hoot about the experts. He's the kind of guy you want next to you in the trenches, cracking sick jokes about the enemy.

In his case, the enemy was (and is) the zoological community. Since the 1980s, he's been charging that zoos were out to make money off the condor, breeding them in cages and selling them to the highest bidder. This is a ridiculous and completely unsupportable charge, but Reid keeps making it: "Condors in Taiwan and Kuwait and such. I'm telling you I wouldn't be surprised."

I admire Les and Sally Reid. Like the McMillan brothers, they were antidotes to claims that the environmental movement was an upper-class plot to steal the people's land. When extractive industries tried to use that argument to get at the condor's habitat, Les and a few of his pipe-fitter friends were happy to tell them that they had no idea what they were talking about.

The Reids retired to a modest home on the outskirts of a small California community called Pine Mountain Club in the early 1980s—a simple, sturdy A-frame with a wonderful view and a bedroom built into the rafters. They were living there when the last wild condor was trapped in 1987, and when the first group of zoo-bred birds was released 1992. Sally and Les didn't bother to go looking for the zoo-bred birds, because to them, they weren't really condors.

Sally Reid was diagnosed with Parkinson's disease a few years after that. When her husband got too old to care for her, he was forced to move her to a managed care facility in Bakersfield, several hours' drive away. When Sally started fading in the late 1990s, Les often drove out to Bakersfield to see her several mornings a week. Usually he turned around and drove back to the A-frame in the afternoon. That's what he was doing when I parked my car in the driveway of his home in March 2002.

On the far side of this mountain, to the south, the greater Los Angeles metropolitan area was dealing with yet another warm, smoggy day. But on Reid's side it was snowing. When nobody came to his door, I took a walk in the snow, marveling at the size of the thick wet flakes. Then I got back into the car and fell asleep listening to a Neil Young album called *Zuma*. The bass solo at the start of the song "Cortez the Killer" always makes me think of condors rising in the wind.

When I woke up, a wiry old gray-haired man in work clothes was banging on the window of the car. "You're that reporter fella? Well come on in then. Let's start talkin!"

Reid unrolled his story as he unlocked his front door. It started in the early 1990s, after he'd unlocked this very same door and walked into his house. Right away he knew there was something wrong. After adjusting his hearing aid, he heard a bunch of ripping and bumping noises coming from behind the bedroom door, up at the top of the stairs.

"I thought it was the cat," said Reid. "Then I saw the cat right in front of me. That's when I went up the stairs as quiet as I could, pulled the door open a crack, and peeked inside."

Eight black birds with leathery heads and white triangles under their wings were staring back at Reid—eight young California condors, all on his bed. While Reid had been out visiting his wife they'd ripped a hole in the screen door connected to a small deck off the bedroom, pushing their way in and hopping up onto the bed. After ripping at the mattress and the sheets for a while, the birds froze when the door moved. One had a chunk of mattress hanging off its beak. Another appeared to be eating a pair of Reid's underpants.

"I said, 'Okay, boys, you're not supposed to be here. You're going to have to leave.' They just stood there staring back at me for a minute or two. Then they turned around and went back out through the hole in the screen. One by one, like a bunch of kids, without any argument at all. I closed the glass door and before long they had their beaks against it, like, 'Hey! Why can't we come in here?' They were lucky it was me."

The birds inspected Les Reid's deck. They walked back and forth on the railing. After a few hours they flew away. Then they came back. This is how it went for a couple of years, and Reid was grateful. He liked it when the young birds opened their wings to take in sunny days, basking and moving only their heads. He liked it when they stomped around on his roof, on the railings. He watched them through a sliding glass door for hours, making up names for individual birds. It made him feel young.

"They had this thing they did with the umbrella," he said. "I've got this big umbrella at one end of the deck and the condors kept trying to stand on it. One flies off the roof and lands on top of the umbrella, but it's too slippery and the condor falls off onto the deck, and then another bird tries to do the same thing and falls off just like the first one. They were having fun out there."

Les Reid knows he should have chased the condors off his deck, but he didn't even try. When government biologists came around he told them to get lost. Then he posted NO TRESPASSING signs all over his property. "The Department of Fish and Game would call and ask me to tell them what numbers were on the birds' tags and I'd say, 'No, I'm not going to do that, it's none of your goddamned business what the numbers are.' And hang up. They'd call back and say, 'We'll get you for that, Reid,' and I'd say, 'No you won't. That'd be the biggest story the local papers ever had.' "

Why the condors picked this particular house is not completely clear. Part of the attraction had to be the sweeping view of the mountains to the west and the north. Another part may have been a steady supply of raw meat from the grocery store. Government biologists have repeatedly charged that Reid was feeding the condors, noting that meat sales at the local deli spiked when the birds were around.

Reid swore he wasn't feeding anything except himself and the cat. He knew feeding condors was both illegal and a stupid thing to do. Anyway, people who did it ran the risk of seeing chunks of their own flesh bitten off, and the additional risk of being arrested and fined. Condors also pay a price when they start homing in on heaping plates of ground sirloin: they lose the desire to act like wild birds. Why spend the whole day flying around when you know the old guy with the view is throwing meat on his back porch? Why not land on that unattended picnic table and scarf down the bucket of fried chicken?

"I would never feed them," Reid said again, looking more than slightly pissed off. "And don't go blaming people like me for messing up the condors. They were defective when they left the zoos. Okay?"

I don't know exactly what Les Reid meant when he used that word "defective," and I don't think he did, either. But when the first of several different groups of condors settled in on his deck in 1998, it was clear that the zoos and the field biologists, not to mention the birds themselves, had a very serious problem. Condors released near the Sespe had been buzzing passing cars, slicing open garbage cans, walking through crowded business districts, and generally acting like a gang of bored punks.

The birds released in Arizona were having problems of their own, frequenting campsites and cluttering up the entrance to an old uranium mine. Shawn Farry was repeatedly forced to flush recalcitrant birds out of Fredonia and Kanab, chasing them past the homes of unemployed people who blamed endangered species for their problems.

Those are the stories I was able to confirm; the rumors were even better: A condor lands in a small town in Southern California and walks into a bank, terrifying tellers and drawing police. A condor in northern Arizona eats a sandwich in the front seat of a Park Service pickup truck and then poses for a picture. A condor at the Hualapai Reservation in Arizona lingers near an airport runway until it is locked inside the pilot's lounge; by the time a field biologist from the Peregrine Fund arrives on the scene, the pilot's lounge has been destroyed and the condor is standing on a chair staring at a television tuned to NBC, which was showing pictures of the war in Kosovo. As far as I can tell, those first two stories are completely false, even though one of them came out of the mouth of a so-called media specialist hired by the Fish and Wildlife Service. But the third one's true, Farry says.

Farry thought reporters asked him way too many questions about the condor hijinks. Every time we talked, he stressed that many of the birds seemed wild from the start. Those were the condors that never slept on ledges the coyotes could reach, and the ones that never seemed to lose the wind. Farry didn't have to climb down and get those birds out of the bottoms of windless canyons, and he never had to jump the metal guardrail and chase them off the boulders in front of the El Tovar Hotel and Lounge on the South Rim of the Grand Canyon. That was also a stupid thing to do, given that a slip could easily send him bouncing down a two-mile cliff. Crowds of gaping tourists often gathered while he worked. Usually they rooted for the birds.

Farry paused to explain why the friendly condors were the most at risk. Birds that flew toward humans and the things they built were probably the first to be "accidentally" shot by folks who didn't recognize them, and among the first to suffer when the human world jumped up in front of them. Out in California, one of these condors died after slurping up what was probably antifreeze. And for a while the zoo-bred birds seemed dangerously fond of electric power poles. The view from the tops of these poles was very good, but coming and going was a problem—condors that ignored or did not see the lines kept slamming right into them and killing themselves. One bird almost sliced a wing off; another was nearly decapitated. One hit the positive power line with the tip of an outstretched wing, causing the bird to flip up and over and down onto the rest of the lines. When the other wing hit the negative current, the body of the condor shook violently for a moment. The accident caused the power to go out in the town of Fillmore, in Ventura County.[1]

Some of the people in uniforms hated doing this work. The man who seemed to hate it most was Dave Clendenen, the lead biologist in the Sespe Condor Sanctuary. Clendenen had been with the condor program since the mid-1980s, ruining his back on those endless zombie patrols and fighting on Noel Snyder's side in the bitter interoffice battles. By the late 1990s, he had spent more time in the field that any other biologist, including Carl Koford. He would do anything to help these birds.

But not this. This was absurd. Clendenen said he felt like a fool when he tried to explain "aversion therapy" to tourists or neighbors.

"I didn't join the Fish and Wildlife Service so I could throw sticks at birds," he said. "Especially when all they did was wait for you to go away. It was obvious to me that we weren't getting very far training condors to act like wild birds, because they weren't wild birds anymore. They weren't the same condors I had known when they were wild."

Clendenen helped restore the California condors in 1992. A few months later he helped capture the condors that were still alive so they could be sent back to the zoo. Then he helped release a second group, much deeper into the sanctuary. It was some of these birds that showed up on Les Reid's porch. Clendenen called Mike Wallace, now at the San Diego Wild Animal Park, on a more or less regular basis to say that there was something wrong with the birds. Wallace told him to wait. The newly released condors were too young to be interested in breeding, and they didn't have a parent to teach them how to act in the wild, so naturally this transition was going to take a while.

The most consistent troublemakers were the young birds that tended to travel in gangs. These birds weren't interested in pairing off and searching for a starter cave, and like all the zoo-bred condors, they were provided with a steady supply of easy-to-find carcasses. So what did they do all day? Basically, they wandered around playing follow the leader, and the leader was often the least cautious condor in the group. Wallace, then the scientist in charge of the condor reintroduction program, said these young and restless groups of birds did exactly what a pack of young and restless humans would do: "They were like a bunch of teenagers whose parents left town without hiding the keys to the sports car," he said. "They're going to fire that baby up and cruise around town."

Wallace liked this metaphor because it implied that these problems would fade as the birds got older. When they started finding mates and tending fledglings, they'd have no time to mess around. Wallace said as much when exasperated field crews called him for advice. Hang in there and wait a bit, he told them. Give the birds a year or two and they'll repent their foolish ways.

Wallace was aware that this approach would fail if the birds kept making lethal mistakes. To solve that problem, he made some changes at the captive breeding centers. He wanted condors coming out of zoos to know that power lines were very bad, and that people were to be avoided. He didn't think it would be difficult to pound these attitudes into the birds, and so he set to work.

Not long afterward, a fledgling condor was minding its own business in an off-exhibit flight pen at the Los Angeles Zoo, waiting for its daily meal of vitamin-fortified horsemeat, when a hidden door in the plywood fence flew open, and a bunch of angry-looking people rushed in. They chased the condor around the enclosure and then into a corner, yelling insults and waving their arms. Then they grabbed the bird and held it tight. Wallace stuck a needle in the bird's left leg. The condor tried to bite him, but it couldn't move.

Wallace held the needle steady in the vein until the plastic tube was full of blood. He opened the bird's beak and looked into the mouth. He put his thumbs near the condor's eyes and pulled the skin back, looking for injuries and signs of disease. Then, on his signal, the other keepers dropped the bird and ran away, slamming the door behind them.

A few days later, this same bird saw that an attractive new perch had appeared at the other end of its pen—a round, brown wooden pole, tall and not too thick, that looked like it'd been soaked in something black and smudgy. A short horizontal board was attached to the top of the pole. Evenly spaced whitish knobs were attached to the horizontal board. Long thin wires stretched forward from these knobs, disappearing into the far wall.

The bird leaned forward, flapped twice, and landed on the pole. When it landed, an awful shock flashed up through its body. The condor bounced up into the air, hissing, flapping wildly. When it settled back onto the new perch it got another painful jolt and bounced off again.

This was a type of the "aversion therapy" that Clendenen so hated. Wallace launched these programs just before he left the Los Angeles Zoo to run the condor program at the San Diego Wild Animal Park. Basically, he taught the other keepers to abuse the birds without actually hurting them. Sometimes that meant charging into big groups of condors, causing the birds to fall to the ground or bolt and slam into walls. Sometimes it meant catching the birds and stuffing them into crates. Wallace also told his crew to run in every now and then and shake everything the birds tried to roost on. Condors soon to be released were mugged repeatedly.

"You wanted to keep them off-balance psychologically," Wallace said. "If we did the same thing every time, we wouldn't be as scary."

Wallace was encouraged by the early results he got from the hazing and aversion-therapy sessions, and so he asked the field crews to try them in the wild. But when the field crews tried to bother the birds as Wallace had, the condors were not nearly so predictably scared.

━━━━

Summer 1996: Several hundred miles up the coast from Los Angeles, near the top of a mountain carpeted with chaparral, Joe Burnett waited for a group of photographers to finish climbing into a camouflaged blind. Farther down the mountain, reporters and dignitaries waited for the historic moment. When Burnett heard that everyone was finally in position, he pulled a rope that opened a door on the front of wooden release pen; four young condors hatched and reared in zoos flew out over the trees.

These condors were the first to see the mountains near Big Sur in fifty years. Long-dead naturalists had written of condors that dined at the bottom of the stark rock cliffs that rose up out of the Pacific, jostling for position and cleaning out the carcasses of whales and elephant seals. It was said that meals like those were once made possible by great white sharks that gorged themselves on local marine mammal colonies, ravaging the breeding grounds and

leaving the half-eaten dead. Zoo-bred condors didn't have the flying skills needed to make it down to those rocks, and if they did
make it down, they would never get back up. The condors in the
box in front of Burnett would be fed on the carcasses of stillborn
cattle collected from local dairies. Volunteers would drive these carcasses into the wilderness areas, and then chain them to the sides of
hills reshaped by countless herds of cattle.

Burnett had been hired by the Ventana Wilderness Society, a
nonprofit conservation group that had just finished reintroducing
golden eagles to these same mountains. He was then a twenty-
something kid from Virginia who had come to California to work
with bald eagles and ended up more or less living his dream. Keeping company with condors wasn't quite what he had expected, but
that only made it more exciting.

"I'm sitting there waiting for the birds to soar up into the clouds
while everybody oohs and ahhs," said Burnett. "'Oh, the mysterious condor,' that kind of thing. What they do instead is fly straight
over to the blind with the photographers in it and land right on top
of the thing. Then they lean over and stick their heads in the openings we made for the cameras. It was like, 'Hello! You're busted in
there!' I was thinking 'Oh shit, this isn't right.'"

Burnett said he'd been worrying about these birds since the day
they arrived. One of the birds seemed sluggish to him, and there
was something vaguely odd about the other three. He couldn't say
why, but these condors looked a little slow to him, as if they were
waiting to be told exactly what to do.[2]

None of these young condors had ever spent time with an adult
member of their species. They had all been raised by condor puppets attached to the hands of captive breeders; after that, they'd
only seen each other. Burnett called Wallace, who gave him his
standard line: relax and wait.

Burnett said he spent the next several months trying to keep a bunch of "shoelace nibbling" condors out of trouble. He kept them from ending up splattered all over the windshields of the trucks and sports cars flying up and down the Pacific Coast Highway and pulled them out of nasty stretches of dried-out chaparral. At one point, he found them sitting in some trees near a bunch of naked sunbathers at the famous New Age spa called the Esalen Institute: "All I could think was, 'Man I hope they know these people aren't dead yet,'" he said. Burnett said that the owners of the institute were very understanding, even though he was dressed like a commando and armed with a "net gun" that looked like a bazooka. "They asked me to stay in the bushes," he said. "That was fun up until the end, when I saw the birds up on the roof of one of the buildings. I climbed up onto the roof and snuck up on them and fired the net and ran over to the edge of the roof to get them and there were about forty people standing right below us. Most of them had clothes on."

Eventually he threw a fit, calling Wallace to tell him he was sick of cleaning up after puppet-reared birds. Wallace said he'd take another look at how the birds were being raised in the captive breeding programs. The puppet-trained birds were captured and replaced by a group of slightly older condors that had been raised by adult birds. They were the only group of condors in the wild raised exclusively by real birds.

This group stayed out of trouble. When it got around that this was happening, Burnett got a call from the man who had published more peer-reviewed studies of the condor than everyone else combined. Burnett had never talked to Noel Snyder before, but he'd heard all about his exploits in the Sespe Sanctuary in the 1980s. Snyder explained that he'd rekindled his interest in the California Condor Program, partly because of complaints from friends still in the field. At the moment, he was working on a broad review of the recovery program, and he wanted to come and see the Ventana birds.

"I told him, sure, no problem," said Burnett. "He was here for several days. I was pretty sure we had a problem with the puppet birds at that point, but when Noel's paper came out, it seemed like he was trying to close down the program. I thought he really over-did it."

The broad review was published in the August 2000 issue of *Conservation Biology,* a scientific journal. Next to Sndyer's name were the names of several coauthors, including Dave Clendenen, who'd just quit his job in the Sespe Sanctuary, and Vicky Meretsky, a computer-modeling specialist at Indiana University, who'd worked with free-flying condors in the 1980s. After plugging every-thing she knew about the birds into a new computer model, Meretsky had concluded that the most expensive endangered-species-protection program in American history was about to crash and burn. Misbehaving condors were dying in the wild with alarm-ing frequency, she found. "Good" birds were being led astray. The birds had not improved with age, as hoped for, and there wasn't any evidence that change was on the way. Extinction wasn't likely, but if present trends continued, success was out of the question.

"The program had become a stocking operation," Snyder said. "Birds were dying in the wild at the same rate that they were com-ing out of the captive breeding programs. Success was still a possi-bility, but not without major changes."

Snyder and his coauthors charged that many of these problems could be traced back to the puppet-raised birds. According to the paper, the birds weren't fooled by the imitation parent birds—they knew there were human hands inside those beaks, and human heads behind the mirrored windows. These were the birds that headed straight for people when they were released to the wild, the paper surmised. Puppet-reared birds that got in trouble had been led astray, the authors said. For proof, consider the Big Sur program, where birds reared only by live parent-birds were acting like condors were supposed to act.

In interviews, Snyder and his colleagues said the condor program was becoming "a perpetual and very expensive black hole." Reversing the trend meant rebooting the entire program. All of the condors in the wild needed to be captured and returned to the zoos. After that, more birds could be released. Since there was no hope for the puppet-reared birds, they'd have to stay in the zoos for the rest of their lives.

Snyder's findings came as no surprise. He aired his concerns at what I've heard described as an "excruciatingly tense" meeting of the condor recovery team, and in the book he had just published. But the harshness of the paper knocked the program for a loop, which was quickly followed by an uproar. People with the program questioned the strength of Meretsky's modeling work and the accuracy of some of her key data. Snyder and Wallace, once close friends, are now said to be the bitterest of enemies.[3]

When I met with Wallace in San Diego in March 2002, he tried to give Snyder his due. "I think the recovery program is in better shape because of the paper and the book," he said. "I don't agree with the conclusions, but we needed a kick in the pants. We made a lot of changes at the captive breeding center after Noel started in on us. We would have made the changes anyway, but not so fast."

The changes Wallace made aren't anywhere near as drastic as the ones Snyder wanted, but Wallace said they've helped. Young birds are released with older ones, and puppet-reared birds are never released without at least one parent-reared bird. Keepers who work the puppets try much harder to avoid being seen, and the puppets have a new personality.

"When we started doing this, the puppet birds were always very gentle with their young, but that's not what the real parents do. If a chick does something wrong, the parent lets it know. It's like *Pow,* a whack in the head. Gotta teach those little guys some discipline, you know?"

Back in the Ventana Wilderness, Joe Burnett has also seen some changes. He still chases condors out of residential neighborhoods, but these days that's relatively rare. He said that as the birds have aged, they've learned to do some of the things condors did in the old days. When I asked him what he meant, he drove me down to the edge of the cliffs.

"See that beach down there?" he said. "About a year ago we stopped getting signals on the radio tags on a couple of puppet-reared condors. We drove around for a long time trying to pick up a signal, and then we got this tiny little blip while we were driving past these cliffs. I make it out to where we're standing now and there they are at the bottom of the cliff, hammering away at the carcass of a sea lion. I saw the giant wings and a lot of blood and it was amazing. I mean this was something condors weren't supposed to be able to do anymore."

It was at just about this time that Les Reid's condors stopped coming around. Reid said he was glad they'd finally figured out which end was supposed to be up. But I think he missed them dearly.

fifteen

THE REAL KILLERS

===

Thy life's a miracle.

—William Shakespeare, *King Lear*

Heavy winds rumble through the empty grazing lands near the tiny town of Gorman, north of where the eight-lane freeway known as Interstate 5 crashes through the once-forbidding Tejon Pass. When the winds are at their worst, they rattle the cars rolling down the part of the interstate known as the Grapevine, forcing cautious drivers to pull over to the side and wait for the weather to die down. Incautious drivers are sometimes punished by gusts that flick their cars off the side of I-5, or into the fleets of tractor-trailers roaring past the rocks that used to hide the bandits and the Indians.

Atmospheric restlessness is common in the mountains at the base of the San Joaquin Valley. In certain places the winds never seem to stop. Condors know exactly where these places are and how they're linked together—they know how to ride them from Ventura County east to the crucial feeding grounds, and then on to the southern end of the Sierra Nevada. If it takes a long time to find a meal, the birds don't have to worry—since they've got an easy ride home, they're free to spend some extra time patrolling.

The winds near the town of Gorman are a crucial link in this chain. On maps of the condors' range they sit at the center of the wishbone, with the coastal mountains on one side, the Sierra

Nevada on the other, and the Transverse Ranges at the base. Condors won't be cut off if the chain is broken, but they will have to use more energy to move across the base of the wishbone.

That's why alarms went off in the offices of the National Audubon Society when a well-connected energy company tried to build a wind farm near Gorman in 1999. Fifty-three towers with two-hundred-foot propellers were to be raised and strung across the birds' flight path, raising fears that condors would be shredded if they failed to see the fans. Activists imagined finding hacked-off wings and giant feathers at the base of the towers. This had never happened to a condor passing over an existing wind farm, but it had happened to golden eagles, and that seemed close enough.

Condors have been killed many different ways over the years—they've been poisoned, shot, starved, stoned, clawed, drowned, beaten, pierced, smothered, eaten, skinned, and even buried alive. In the 1920s, three condors died when they tried to fly through a hailstorm. But of all the sordid ends these birds have managed to meet, none compared with being pureed. That was the opinion of Dan Beard, the Audubon staffer whose job it was to keep the towers from being built.

The energy company that wanted to build them was called Enron. Through the Clinton years, it had worked hard to cultivate a "good guy" image with environmentalists, mostly by proclaiming an intention to invest in alternative energy sources. That image would eventually be vaporized a few years later when Enron turned out to be a corrupt corporate hellhole. But in 1999, the company was the environmental golden child, and golden children don't kill condors. They build solar energy arrays and energy-saving freezers, and nonpolluting wind farms that resemble modern art. Big metal towers, all in rows, all with spinning propellers. As if Christo himself had installed it. Spokesmen for the company were said to be insulted by the thought that they would try to build anything like a "condor death trap."[1]

But the National Audubon Society said that that was exactly what was happening, at a press conference held just after the Enron Wind Corporation unveiled its plans. "It's hard to imagine a worse idea than putting a condor Cuisinart next door to a critical condor habitat," said Beard. "We believe this project should be stopped." Beard made it clear that he didn't want a smaller wind farm built, or a perhaps a wind farm where the towers were "improved" in ways that might have made them easier to miss. He wanted to kill any idea of wind farms near Gorman.

Audubon had derailed a plan to build a larger set of towers in this same location back in 1989, but that was not a project backed by Enron. Wind power was then the world's fastest growing source of energy, and many environmentalists were praying for continued growth. These activists stressed that wind farms were clean sources of electricity, producing no radioactive waste and virtually no pollution. Bill Clinton's Energy Department was subsidizing wind farms by handing out 5 percent tax breaks, and California seemed to be the market of the future.

Beard's response to all of this was adamant and politically awkward. He said Audubon supported the idea of wind power, but not when it posed a threat to birds and especially not when it threatened to mince a lot of condors. The problem with this caveat was that birds and wind farms often need the same kinds of winds: that's why lines of turbines tend to end up on the ridgelines of mountains in the middle of migratory pathways. And for a while the talk-show hosts and columnists circled like sharks. Depending on whom you listened to, this wind farm was a reason to drill for oil and build nuclear plants—proof that environmentalists were people to be ignored—or a great reason to invest in solar power and hydrogen.

Meanwhile, Enron was threatening to bury Audubon. Publicly, the company kept saying it was sure the towers wouldn't hurt the condors, but in private meetings, Enron took a much harder line. It was going to use its power and influence to make the activists look like fools and then it was going to go ahead with the wind project. Enron was then thought to have enough money to stick around until its enemies had to give up, and the public-relations skills to make it look as if it had taken the high ground. If Beard and his colleagues didn't step aside, they would look like friends of the big polluters they'd spent so many years fighting, and the condor would be painted as an anti-environmental bird.

Beard and his colleagues ended this game by taking a page out of David Brower's playbook, and with a semisecret deal that Brower never would have made. Full-page ads denouncing Enron ran in several newspapers, and large amounts of billboard space were purchased in Los Angeles and in Enron's home city of Houston. The billboards showed a giant vulture headed toward the turbines, over the words ENRON WIND CORP. and the words KILL THE CONDORS?

The semisecret deal aligned Audubon with the owners of the Tejon Ranch, who were then laying plans to develop large parts of the California condor's range. The owners of the ranch thought the wind-farm plan might get in the way of their development plans, partly because the "visible blight" of the towers would be close to the freeway. After paying for the billboards and the newspaper ads, the ranch owners went to Enron and offered them a deal: give us that land and we'll give you some land that is out of the condors' flight path.

When Enron took the offer, Beard called the press and praised it as visionary. "Enron Wind has a long history of dealing with environmental concerns in a positive, responsible manner," Beard

said. "Enron Wind has clearly proven that it is a company committed to protecting the environment." Audubon also threw its uncon-
ditional support behind a push to reauthorize the Wind Energy
Protection Tax Credit Act, which had helped make it possible for
Enron to purchase the land near Gorman in the first place.

That was when the owners of the Tejon Ranch unveiled a plan to
build a big residential development on the former site of the wind
farm. "We got screwed," said a former member of the California
Condor Recovery team. "It was as simple as that." Actually, it wasn't
as simple as that, since the ranch had made no promises, but relations between the ranch and the condors' keepers had been raw for
years by then, and it was easy for each group to assume the worst of
the other.

The owners of the Tejon Ranch had been trying to get their land
out from under the condor since the early 1970s, when a huge part
of the ranch was identified as critically important to the future of
the species. Technically, the "Critical Habitat" rule in the Endangered Species Act doesn't have a lot of legal force, but the fear that it
might, if tested in a court of law, is the kind of thing that makes
banks withhold loans. Lawyers representing the ranch tried repeatedly to work paragraphs into recovery plans that would have freed
them of any obligation to accommodate the condor. This strategy
appeared to peak in 1998, when the company argued that the condor had been "technically extinct," as the last wild bird had been
captured in 1987. That argument was never made in court, but the
chance that it could be lingered for years, until the warring parties
cut a deal.

In March 2002, the captive breeding programs reached a milestone.
After falling to a probable low of fewer than twenty-five birds in the

1970s, the species had now rebounded to two hundred condors alive. Educated guessers said it had been at least a century since there had been so many condors.

One of the condors that had stopped laying eggs, AC-8, also known as the Matriarch because of her age (thirty-nine) and because she'd laid so many eggs at the captive breeding center at the Los Angeles Zoo, was to be released in the Sespe Wilderness to act as a "mentor bird" to zoo-bred condors struggling in the wild. Joe Burnett had cooked up the plan with the help of Mike Clark, a condor keeper at the Los Angeles Zoo who'd worked with AC-8 for years.

But Noel Snyder didn't like the plan at all. He'd known the Matriarch as she'd grown up in the wild and paired off with Igor. Snyder and his colleagues on the zombie patrol had taken her eggs and flown them down to the captive breeding center. But it wasn't his personal attachment to AC-8 that drove Snyder's opposition to the so-called mentor plan; he just didn't think any of the birds were safe in the wild. "It stuns me to think that we're releasing birds into such a poisonous habitat," he said. "That was one of the reasons we removed them all in the 1980s. The galling thing is that we now know what the poison is and where it's coming from, but nobody has been able to do anything about it."

Snyder says the problem is lead shot in the carcasses of animals left behind by hunters. When condors eat the carcasses they swallow lead pellets and lead-bullet fragments, and lead is just about the worst thing you can put into a condor. It's a toxin and an element, which means it never stops being poisonous. Lead also binds to common proteins in the bloodstreams of animals, so that it can't be flushed out by the liver. Lead ruins the brain and the central nervous system, but it usually kills condors by closing down the nerves that work the digestive system.[2]

Condors that eat a lot of lead die slow, horrible deaths. Near the end, they can barely crawl. Low doses of lead may degrade a condor's flying skills, raising the odds that the bird will be killed by an eagle or a coyote. Other condors crowd in and push such a bird off carcasses, just because they can. Reproductive problems are another side effect.

Shawn Farry told me all about the single biggest lead scare in the history of the program. It started on June 6, 1999, when a condor in the Grand Canyon ate what may have been its final meal. With the help of twelve friends, condor number 165 peeled the meat off a carcass in some bushes near the edge of the national park.

A few days later, 165 disappeared. Farry stopped getting signals from the beepers on the bird's wings, which either meant the condor was on a long trip or that it was now dead. Dying condors sometimes crawl into caves that block the signals from radio-tracking devices on their wings, and when they end up at the bottoms of canyons it is very hard to get a reading. Farry and the rest of the field crew launched an all-out search that ended when he saw the bird lying on its back at the bottom of a canyon near the big resort hotels. The right wing, half open, was caught in a branch; the left was crushed against the body. The partially buried head pointed toward the top of the 45-degree talus slope. The legs were sticking straight up into the air. Farry made his way down to the bird and pulled out a notebook:

The carcass, specifically the top of the head and the leading edges of the folding wings, are covered in fine dirt consistent with the surrounding soil. A likely impact site is located approximately twenty feet upslope of the carcass, along with a slide path to its final location. The contact of the left wing with the dead branch appears to have arrested the carcass's slide. Small fragments of vegetation and fine debris cover the carcass along with bits of sharp vegetation "puncturing" flight feathers further indicating an impact. The location also indicates a direct "dead" fall from the rim above with no indications of a controlled glide or descent.

Judging from this evidence, 165 had fallen off a cliff and plummeted straight down to the bottom of the canyon. Farry saw nothing that would indicate an attempt at flight. This bird was dead or very nearly dead before it hit the ground.

Farry found no bullet wounds or signs of an attack. The left wing was broken, but that had almost certainly happened after the bird fell. Number 165 historically had been a very healthy bird, and a good flier: "Nothing in this bird's history foreshadowed this mortality."

The carcass was flown to a zoo where pathologists immediately found the cause of death. Seventeen lead pellets showed up in the X-rays, all of them in the gut.

A few days later, field biologist Gretchen Druliner noticed that condor number 91 had given up eating and crawled under a boulder near the Vermilion Cliffs. The bird was alive when she got to it, but only barely. Druliner brought it to the office at Vermilion Cliffs.

"At this time [the condor] could barely stand or keep her eyes open and weighed only ten pounds," Farry wrote. "She appeared to be approximately ten percent dehydrated and severely emaciated. A blood sample was taken and we administered 180 cc of subcutaneous fluids. She also drank approximately two cups of water on her own."

The bird got better, then crashed. Farry drove it to the Phoenix Zoo, where a blood test showed severe anemia. A blood transfusion helped, but only briefly; after perking up a bit, 91 fell over dead.

Condor number 82 was the next to go: a biologist found it on the ground a little farther from the Vermilion Cliffs. "Face down, wings open," Farry wrote. "The carcass was in extremely poor condition, indicating that she had been at this location for some time." Farry's hunch was that the bird was trying to return to the Vermilion Cliffs when its wings gave out. After the bird hit the ground and died, coyotes tore it to bits.

Traps for the remaining condors had been set by then—big mesh boxes over the carcasses, with an open door at one end. When six or seven condors had walked through the door, a field biologist would snap it by pulling on a hidden line.

At first the condors kept their distance, which came as no surprise. "They'd all done the same thing before and ended up in kennels." These were remarkably wily birds, but Farry was too rushed to be impressed; for all he knew, every condor in the area had lead pellets melting in its guts.

"I'd sit there in the blind and watch them land on top of the trap, and then walk around it, and then jump back on top, and I'm thinking, *Please go inside, please, please, please.* But they don't."

Another bird was AWOL by this time: condor number 150 had disappeared after flying into Marble Canyon. A very weak set of beeps seemed to be coming from a section of the Colorado River called the Sheer Wall Rapids.

"Attempts to triangulate the signal by circling around to the eastern rim [from the western rim] failed," Farry wrote. Climbing several thousand feet up to the top of a place called Echo Cliffs didn't help—Farry couldn't find the signal up there, either. The Hatch Company lent a hand by allowing him to stick telemetry devices to one of their planes, but this attempt also came up empty. "The signal remained stationary, extremely weak, and defied pinpointing," Farry wrote in his notes. Condor 150 was the first condor ever hatched at the Peregrine Fund and was never found.

Farry and the crew lured the remaining eleven condors into the walk-in trap with a carcass inside it. The newly captured birds were locked in kennels and driven to the trailer at Vermilion Cliffs. Inside, Farry and his colleagues pulled the birds out of the kennels and held them down on a table. Blood was drawn and run through a brand-new high-speed lead-testing machine. Three minutes later, a rough lead count flashed on a screen, sometimes next to the word "high." Birds with high counts were loaded back into their kennels and driven over to the veterinary clinic in Page. If X-rays there showed pellets, the condors were loaded up again and hauled to the Phoenix Zoo.

The pressure was overwhelming. "I'd be standing there with five or six kennels with condors inside them lined up on the kitchen floor," Farry recalled. "We'd take a bird out of the kennel and draw some blood and spend the next three minutes totally freaked out. Then the machine beeps and the guy next to me just says 'high.' When he says it for the fifth time in a row, you know it's going to be a long night."

Some of the birds in the kennels were barely breathing; others were bouncing around and screeching like they'd been stabbed. One of the first lead readings seemed impossible—condor 158, a five-year-old male, had the highest lead count ever recorded in a California condor. "Three hundred ninety," Farry said. "This bird should have been dead. A count in the mid–three hundreds was supposed to be lethal. If anybody ever needed proof that these birds were resilient, this was it."

Condor 158 had six lead shotgun pellets in her gizzard. Condor 133 had one in her intestine. Condor 136 had two in the gut; condor 119 had one.

These birds were driven to Phoenix to have the pellets removed. Other birds were forced to endure an extremely painful cleansing process called chelation therapy. One or two people would hold a bird while Farry injected a chemical called calcium EDTA into its chest muscles. Each bird needed two injections a day for at least five days. The shots were extremely painful, even when the calcium EDTA was mixed with painkillers.

Farry said he thought about how much the birds were going to hate the sight of his face for the rest of their lives. "We'd been trying to teach the birds to stay away from people, and one of the ways you do that is to make the birds associate humans with pain. It didn't seem like they were going to have problems making that association after this was over with."

The first round of chelation shots did what it was supposed to do—lead levels in all of the birds started to fall. A second round of shots pulled the levels down even further. Except for 191—the second bird to die—all nine of the captured condors were now out of danger. By putting the birds and the field crew through several weeks of hell, Farry had saved the reputation of the Peregrine Fund. If all or most of the captured birds had died, it's likely the Arizona reintroduction program would have died with it.

But what was he supposed to do now? Rerelease the condors back into the environment that had just poisoned them? Send them to the zoos and wait for a new set of birds? Farry and his colleagues tried to think the issue through. The first problem was that they didn't know where the lead in the birds had come from, and there were a lot of possibilities.

So what were the clues? According to Farry, one was that these condors liked company. They followed each other around and ate from the same carcasses, like the one he'd seen them eating from just before the crisis developed. Another clue was that the birds all got sick at once. Farry thought the clues pointed toward a "massive freak event," which was not likely to recur. All he had to do was find a giant, half-eaten, lead-filled carcass in the vicinity and the case would have been closed.[3]

But he didn't find it, and there was another clue that didn't seem to fit this explanation. Lead pellets pulled from the birds were of different sizes, which meant they were almost certainly fired out of several shotguns. This raised several possibilities, two of which would have been alarming. If the condors got the lead poisoning by eating carcasses in several different locations, it would mean that no gunshot carcass was safe for the birds and it would be time to send them back to the zoos. The other scary thought was that a group that didn't like the condors had prepared a booby trap, filling one carcass with bullets. Farry thought it far more likely that some small-time rabbit hunters did it by mistake, by stacking up several dozen rabbit carcasses and posing for a picture. Or maybe it was a bunch of drunk teenagers opening up on a dead horse. In the absence of the carcass, there was cover for all kinds of theories.

Farry decided to write the episode off as a well-meaning mistake. "Maybe someone who loved condors thought he would make sure they had enough to eat, and so he shot a bunch of animals and laid them out where he thought the birds would find them."

Farry released the birds in the fall. Over the years, this process had become routine. But not this time. This time Farry was worried that he might be making a terrible mistake. That's what he says he was thinking about as he prepared the birds to be released. First he tested new radio transmitters on the wings of the birds, attaching them with "cable ties, nuts and bolts, dental floss and super glue." Then he weighed the birds, checked their vital signs, and double-checked the transmitters.

"Then the condor is carried toward the cliff edge and gently released on the ground," he wrote in his notes. "Typically the condor takes a few seconds to regain its composure before it takes to the air, catching the strong uplifting winds generated by the sheer sandstone walls." Farry and his colleagues had seen that happen so often they sometimes took it for granted, but this was not one of those times. "On October 26, 2000, releasing condors back into the wilds very different," he wrote. "To say we were a bit apprehensive is an understatement."

sixteen

ELVIS REENTERS
THE BUILDING

====

Mike Wallace started thinking about hang gliding in the early 1980s, when he was working with Andean condors in Peru. When he saw California condors shadowing gliders near the Sespe in the 1990s, he knew he had to get up there. "I wasn't *that* old," Wallace said. "Anyway I talked about it and talked about it until my girlfriend cut a hang glider ad out of the paper and slapped it down on the table in front of me. 'Okay, Mike, no more excuses.'"

We talked while driving north along the foothills of the San Gabriel Mountains on a ten-lane interstate freeway. I sat next to Wallace with a tape recorder and a notebook in my lap. The back of the pickup we were riding in held all of the essential field supplies—doughnuts and candy bars, a framed picture of Carl Koford, a backpack full of camping stuff, a garbage bag with dirty laundry inside it, one running shoe, and a silver briefcase containing a digital camera with a giant telephoto lens.

We were going to a condor party on the old Hopper Ranch, now the Hopper Mountain Refuge. Wallace was to be there as a representative of the captive breeding program at the San Diego Wild Animal Park and as the chairman of a panel that's supposed to advise the condor recovery program. I'd come out to talk to him at the wild animal park, and then at his home in a rural corner of San Diego County. The house he lived in with his wife and daughter was surrounded by an avocado orchard, and Wallace had dabbled in the sales end of that business for a while. But Wallace didn't need such cutthroat competition, and it wasn't long before he gave up trying to make a profit.

"Here's the thing about hang gliding," he said while pulling out to pass a gargantuan truck. "Every time I do it I am overwhelmed by the condors' flying skills. They can see a black patch of ground a mile away and instantly know the exact size and shape of the thermal wind rising out of it. As a nonflapping entity, I am always trying to calculate these things, and it always amazes me to think that the birds do it naturally."

"Nonflapping entity?" I asked.

"Airplane, helicopter, rocket ship, blimp—pretty much everything that flies that's not a bird or a bat."

The Foothill Freeway merged with the westbound Simi Valley Freeway, which rose and banked to the west. Elevated interstates are no longer built in California—they fall down in earthquakes—and in a way that's sad. I like the way they force people to look out on the worlds they are missing as they drive back and forth from work.

"Look at that line of thermal heads," Wallace said, pointing at a line of clouds that made me think of bouffant hairdos. "Underneath each of those clouds is a powerful column of rising air. When they line up like that, it's called a cloud street. Soaring birds and really good hang glider pilots use clouds like those as stepping-stones."

"What if you're a bad hang glider pilot?"

"Then you get sucked up into the cloud or stranded out in the middle of the San Fernando Valley. You land in a parking lot if you're lucky. If you're not, you crash into a building."

The sign for the Golden State Freeway exit said 2.5 miles. Has Wallace ever flown with the condors? "No," he said, wistfully. "I can't do that. I'm the guy who ends up on the ground with the radio, telling other pilots not to get too close. That doesn't help if the condors *want* to get close, which happens frequently. But no, I don't think I'm ever going to live the dream of looking over and seeing a condor arcing across the sky right next to me. I've done that with seagulls, red-tailed hawks, golden eagles. But not condors."

We leaned into the off-ramp to the Golden State (better known as the I-5) north and swirled through a corkscrew 275-degree turn. After flying forward for a couple of minutes, we swirled in the opposite direction, landing on the westbound side of State Highway 126. The road my family knew as Blood Alley was now a four-lane thruway to the ocean, and not the hair-raising thrill ride down the north side of the Santa Clara River. South of the river were the familiar lines of orange trees, but not for very much longer: the foreign conglomerate that had just bought the Newhall Land and Farming Company was planning to drop several hundred houses on top of them. Environmental groups were trying to persuade the conglomerate to rethink its plans, but it had refused to give an inch, and the Los Angeles County Board of Supervisors was ready to approve the project. It would be the largest single suburban development ever approved by the county.

We passed the turnoff to the town once known as Mr. Cook's Garden of Eden: Piru City, Ventura County, population 356. As we passed, I looked to the hills behind the town, where Mr. Cook's mansion had once stood. It had been reduced to ashes in 1982 when a painter dropped a torch on the roof. Ventura County firemen raced to the scene, but they couldn't find a fire hydrant. The trucks ran out of water as the mansion went up in flames. In the chaos, the firemen failed to notice the pool next to the house. While the mansion was burning, a television reporter asked Scott Newhall, "How does it feel?" "Wonderful," Scott replied. "Really *&^&%$#@ great! How the hell do you *think* it feels to watch your house burn down!" The footage never aired.

Scott and Ruth Newhall, family friends and two of my personal heroes, were the owners of the mansion when it burned. They'd bought it in the 1970s when Scott moved out of the executive editor's office at the *San Francisco Chronicle*. When I was in college in Northern California, I used to visit them all the time. Scott told stories from the "wacko years" he spent at the helm of the *Chronicle*. Ruth repeatedly demolished me at Scrabble.

Scott died unexpectedly in 1987. But before he died, he and Ruth built a copy of Mr. Cook's mansion on the charred foundation of the first one. Except for the sprinklers in the new ceilings and the statue of a phoenix on one of the towers, the new house looks exactly like the old one. More precisely, it looks the way the old one must have looked when Mr. Cook built it at the end of the nineteenth century. I can see him standing on the turret at the top of a brand-new red stone tower, gazing out at the future metropolis of Piru City, California.

"Is that place a metaphor or what?" I really did say that as the new mansion came into view. Wallace didn't know what I was talking about and I didn't really want to tell him, so we drove on without saying a word. At Fillmore we swung north toward the birds, following the still-wild Sespe Creek to a sharply angled oil road etched into the side of a cracked, dry mountain. As we rose, we sank into the folds of this mountain, losing sight of everything behind us. Hammer-shaped oil pumps covered with rust bobbed endlessly up and down. For a time the groaning of the pumps seemed to be the only sign of life. But as we neared the top of the oil road, the view changed radically as the sky took on a life of its own, looking bluer and deeper and bigger than I'd ever seen it look before. I could almost see the heavy winds that were pushing the clouds and the birds around. When Wallace stopped the pickup truck to open a gate, I heard the wind passing over us, rattling the leaves on the branches of the trees.

We passed the ruined shack that had been Carl Koford's home in the years before World War II. Just before dark, we reached the battered ranch house that served as a base for the condor field team. I sneaked Koford's portrait photo onto a shelf in the living room, over a fireplace that had just been condemned. Four more condors were to be released near here in the morning. Three were zoo-breds. The other was the wild condor Pete Bloom trapped in April 1987. Good old Igor.

The plan to put Igor back where he had come from was approved in December 2002, at a meeting of the condor recovery team in the education building at the Los Angeles Zoo.[1] The first thing I saw when I entered the room was the big, stuffed gorilla in the glass case. Then I saw the big, stuffed tigers and the big, stuffed grizzly bear. They looked a little glassy-eyed, but that's to be expected when your eyes are made out of glass. I wondered what their stories were and when those stories had ended.

Several dozen people were sitting in rows of uncomfortable folding metal chairs, and more people were on the way in. According to the fat blue binders we were handed at the door, talk of Igor's future would come late that afternoon, after a review of the various substrates used to line the bottoms of portable kennels and before the discussions of "telemetry dilemmas," "Listserv etiquette," and something called "the Nixolite Experience." Bruce Palmer, then the director of the condor recovery team, opened the meeting by announcing that the budget for the coming fiscal year would be $1.7 million. Palmer added that most of the money would be spent to repair aging captive breeding centers at the zoos. "One point seven million dollars is a lot of bucks, but unfortunately it's not enough," he said.

A groan ran through the back of the crowd at that point. The field crews were not happy. They were the ones who were hurt the most by the perennial shortage of funding: eighty-hour work-weeks on temporary contracts with pathetic benefits were more or less par for the course. Palmer didn't like it, either, but there wasn't much he could do: critics of the Endangered Species Act had been hacking away at the budgets for projects like this one for many years. The need to find out what was killing condors in the wild was desperate at this point, but the puny budget made that hard to do.

"I came into this program thinking I knew what was going on," said Palmer when I talked to him later. "Turns out I didn't have a clue. Every time you think you understand this bird it does some-thing nobody expected. As for the politics, you don't want to know. That's all I'm going to say."

Talk of lead and misbehaving birds was much more civil than I thought it would be, partly because Noel Snyder and his allies had been encouraged not to attend. They'd tried to make the case for a ban on lead bullets at the last meeting of the recovery team, but that encounter had produced more heat than light. The state of California had commissioned a study of the lead-poisoning prob-lem, but it wasn't likely to be ready for months. After the meeting, Snyder's strongest ally on the recovery team resigned from the Fish and Wildlife Service to manage a private refuge. Robert Rise-brough, a toxicologist revered by some for the work he'd done to help convince the Environmental Protection Agency to ban DDT, said he didn't think lead bullets were the source of the condor's lead problems. Based on what he called a preliminary study of lead readings from the condor's rangelands in California, Risebrough said the real problem could be polluted rain. Risebrough wondered whether rainwater with lead traces from many sources was collecting on the leaves of plants that were eaten by deer. When those deer were shot, or killed in more "natural" ways, the condors that came down to clean the carcasses would have picked up the accumulated doses. Risebrough thought this theory explained why blood-lead levels in the condors seemed to fall at the height of the hunting season, since that was also when the fall rains would have washed the lead off the leaves.

Risebrough said his theory didn't work for condors living near the Grand Canyon, where fragments of lead were often found in the gullets of dying condors. But he wasn't sure the poisoned birds that had turned up so far in Arizona were victims of a widespread hunting problem. Risebrough thought most of these dead birds had been done in by "a single mass-poisoning event," implying that it almost certainly would not happen a second time. Later he would change his mind about this, but at the meeting I attended, he came across as a hard-line skeptic of Snyder's belief that hunters with lead bullets were the culprits.

Discussions of the misbehaving birds were even briefer, perhaps because no condors had been seen on Les Reid's deck for close to a year. Incremental signs of behavioral progress were reported by all of the field crews, and the consensus seemed to be that the birds were getting wiser as they aged. Condors brought up by puppets did not seem to be raising more hell than the other ones were, which wasn't entirely good, since the puppet birds and the parent birds were continuing to perform in front of cheering crowds of tourists at the El Tovar Hotel in the Grand Canyon.

The Igor discussion went by in a hurry, even though many people in the room seemed to have a lot to say about him. Behaviorists were worried that Igor might butt in on well-established pairs. Others said that didn't matter, as Igor was a well-known stud. Someone wondered why the zoos would want to give away a well-known stud; the answer was that Igor's genes were so well represented that his services were no longer needed.

In the wild, Igor would probably settle down with AC-8, the Matriarch, his old breeding partner: she had been the first condor taken from the wild to be released, a year before. AC-8 had also been a breeding machine, but she'd recently had a tumor removed, and veterinarians wondered if she'd become sterile as a result.

"Releasing AC-9 would make a pretty good movie of the week," said Shawn Farry. "But is this more than a stunt? Maybe it's a good idea to put an older bird back into the wild with the young ones, but why does it have to be this one?"

The unspoken answer to Farry's question seemed to be "Why not?" as news that Igor was to be released could bring reporters by the dozens. But the trapdoor in the bottom of that argument was opened by botanist Maeton Freel of the Forest Service. Freel had spent a good part of his long career working near the condor refuge, where he'd watched Igor dodge the proverbial bullet many times. Freel remembered the dirges that ran in the papers when Igor was captured and taken to the zoo, and he knew there was a chance that this maneuver might not have a happy ending.

"What if he dies?" Freel said, silencing the muttering in the back of the room. "AC-9 has now spent more than half his life in captivity, and we really don't know what it's done to him. There's no way to predict what he will do in the Sespe for the first time in fifteen years."

There was only one point on which everybody in the room

agreed. If Igor did go home, he'd find it hard to lose those pesky human trackers. Field crews in the Sespe had just finished testing a tracking system built around a thirty-five-gram satellite transmitter, which had been attached to the right wing of the Matriarch. She was still attached to a set of old-fashioned transmitters, but the satellite beeper had been putting it to shame. Condors wearing the satellite transmitters never flew out of tracking range; anytime anybody wanted to know where a satellite bird had been, all they had to do was turn on the computer and download the latest tracking map. The bird with the thirty-five-gram device would appear as a blip at the end of a line that looked like it belonged on an Etch-A-Sketch: it would be a record of every flight the bird had made. In the past, the only way to make these maps was to trap the birds and download the data from their transmitters; this was a change that seemed to make it safer *and* easier to study the tagged birds. But as anyone who's ever worked with condors understands, these birds always zig just when you think they're going to zag.

———

Mike Clark, the condor keeper at the Los Angeles Zoo, carried a portable dog kennel into Igor's pen on the morning of February 4, 2002. Not long afterward, a helicopter carried Clark and his cargo past the San Fernando Valley and the south edge of Topa Topa Mountain, touching down near the edge of the Sespe Condor Sanctuary. Clark hopped out and carried Igor's kennel into a wood-and-wire holding pen. Three young condors raised in zoos were already there. At first the zoo-breds kept their distance from Igor, knowing he would show them who was boss. By the end of the day, they were following him and trying to copy his movements.

Two weeks later, the four condors were moved to a larger pen in a valley in the Hopper Mountain Refuge. Field biologists and vol-

unteers grabbed the birds, bolting on the ID tags and radio transmitters. When they finished, Igor was pumped.

"He was flying all around the pen," said a volunteer named
Anthony Pietro. "Then he was up at the top of the pen, holding on
to it with his beak. For a minute we thought he was going to break
his beak. He was totally going berserk."

Just before dawn on the following morning, the birds were
moved to a release pen on the lip of a cliff in the center of the sanctuary. Through a wire door on the front of the pen the birds looked
across a canyon at Koford's O.P., one of the wildest places left on
Earth. They saw the hills cracked open by the earthquakes and the
washes carved by prehistoric floods. Much of the chaparral had
never been cut and never would be.

Then there were the cliffs and the caves. This was a landscape defined by cliffs of almost every shape and size, and by an even wider
range of caves. Some were huge and easy to reach. Others were slits
in polished walls. There were caves whose walls were lined with
whitewash sprayed by long-dead condors, and caves that had never
been used. Counting them was out of the question, I thought. It
would be like counting the stars.

Tony Pietro was watching the birds through a peephole in the
back of the release pen. Now it was the zoo-breds that were skitter-
ing around and gnawing at the inside of the pen; Igor had walked
straight from his kennel to the wire mesh door that opened out
onto the Sespe and stood there surveying the landscape. "He
looked to his left, to his right, up, down—he was totally recogniz-
ing it," Pietro said. "I'm not a scientist, but it was like, 'Hey, I remember this place.'"

Igor would have seen the crowds of people gathering on a distant
escarpment, just below Koford's O.P. Chumash dancers, ranchers
and activists, field biologists and breeders, yet no one from the

Peregrine Fund. Someday they will have to explain what they were doing on the day Igor finally came home. The California field crews took their absence as an insult, as did I.

The man Igor would have seen at the outer edge of the escarpment was geneticist Oliver Ryder of the Center for Reproduction of Endangered Species at the San Diego Zoo. He'd never been to the Sespe, and the only condor he'd ever seen outside a zoo was the one perched on the roof of a house. The big guy was Bruce Palmer, the director of the condor recovery team; the people crowding around him were reporters. The Chumash dancers were moving in circles near the point of the escarpment, chanting songs the condor could not hear.

A smaller group had gathered at the top of a rise near the other end of the escarpment. The short guy with the beard was Jon Schmitt. I was standing next to him. Next to me was Pete Bloom, staring through his sighting scope at the front of the release pen and mumbling under his breath. I never had a chance to ask him what he was saying but I think I can guess: Bloom was probably talking to his bird.

The door on the front of the release pen was opened by an unseen hand at 11 A.M. on May 1, 2002. The condors hatched and reared in zoos hopped out right away and flew down to a ledge. But Igor would not leave the doorway. He was still standing in the shadows of the pen when I left the area. Late that afternoon an impatient biologist entered the release pen and nudged Igor out into the world.

The condor known as the Matriarch, Igor's mate in the early 1980s when they were among the last free-flying condors on Earth, was shot and killed on February 13, 2003. The Matriarch was roughly forty when she died. She'd been released in the wild several years before, in the hope that she would teach the misbehaving zoo-breds to act like wild condors. At first she had refused to do any such thing, avoiding the other birds and shooing them away when they came around. The satellite readings showed that she was following her old flight paths to familiar foraging grounds, some of which had lost their carcasses forever in the fourteen years the Matriarch had been away. But in time she'd become a mentor to the younger birds, soaring in their company and showing them her roosts. Trappers had been forced to take her in fall 2000 when blood tests showed that she had a potentially lethal amount of lead in her veins and arteries. When the lead was removed and the bird was re-released in December 2000, you could almost hear the field crews and keepers breathe a sigh of relief.

When Igor was moved to a holding pen in the Sespe just before he was released, the Matriarch had landed on the roof and pulled at the mesh above her old breeding partner. Now, nearly a year later, she was a twisted carcass wedged between a pair of branches near the top of a thirty-foot oak tree on the Tejon Ranch, forty-five minutes north of Hopper Mountain Refuge. A single bullet had passed through her torso and then through her left wing.

Nobody admitted to the crime. Then investigators stood the carcass up on the perch where she'd been killed, tracing the path of the bullet back to where the shooter had likely stood. There they found a rifle shell that led police to the home of a twenty-nine-year-old man who denied firing the shot, even though he had been hunting pigs on the Tejon Ranch that day. After meeting with his lawyer, Britton Cole Lewis changed his mind and admitted that he'd shot the Matriarch. But Lewis swore he didn't know the giant black bird in the oak tree was a California condor. Lewis lived in a town where local hunters would have known that turkey vultures don't have ten-foot wingspans, and the Tejon Ranch says it makes a point of telling visitors not to shoot the giant vultures.

Condor biologists had trouble buying Lewis's story, but federal prosecutors apparently did not. Instead of charging him with a felony violation of the U.S. Endangered Species Act, they accused Lewis of violating the much less powerful Migratory Bird Treaty Act. ESA convictions can mean more jail time and much heavier fines, but they're hard to obtain when prosecutors can't prove the killings were intentional: in other words, when hunters caught dead to rights want to neutralize the ESA, all they have to do is say they pulled the trigger before they knew what they were killing.

Lewis was sentenced to sixty hours probation, a five-thousand-dollar fine, and two hundred hours of community service. At the sentencing he said he was sorry and wished he could take back what he'd done.

———

The last time I saw Igor he was soaring near my old hometown of Piru, at the base of the Topa Topa mountain range in south-central California. I saw him circling slowly around a rising column of wind in the middle of the Sespe Condor Sanctuary. When he reached the top of it, he seemed to pause and scan the wildlands below him. Then he flexed his wings and veered off to the south, toward the smog.

AFTERWORD

S ophie Osborne pulled the garbage can with the carcass of the stillborn calf inside it out of the back of the pickup truck she'd parked near the edge of the Vermilion Cliffs. The can had a pair of shoulder straps attached to the side of it; Osborne grabbed them and swung the can up onto her back. Then she hustled off across a red plateau strewn with ankle-breaking rocks. From behind, it looked a bit like she was skipping.

Osborne is the Peregrine Fund biologist who took Shawn Farry's place in the field, when he quit his job to work with Mike Wallace at the San Diego Wild Animal Park and then quit that job to work with monk seals in the South Pacific. Early in 2001, the first recorded California condor egg was produced in the wild since the last of the free-flying birds were trapped in the 1980s. It was a milestone the recovery team had been praying for since the released zoo-breds had gotten old enough to breed in the wild.

"I noticed that the pair of condors had stopped flying together and settled into a cave," she said. "When one was foraging the other always stayed behind in the cave. When the foraging bird came back to the cave, its mate would fly away. When they alternate like that, it usually means they're guarding an egg."

Osborne began trembling when it dawned on her that she was almost certainly looking at an active condor nest site. News of the event shot through the recovery team like an electric power surge. Then the parent birds appeared to kill the egg. Osborne saw the parent condors fighting near the entrance to the cave one afternoon. One of them went inside and came back out a short time later with the egg impaled on its beak.

One year later, an Arizona breeding pair laid an egg in a cave inside the colossal rock formation called the Battleship in the heart of the Grand Canyon, while at least two other condor eggs were laid in California. A chick emerged from one of the eggs produced in California, but it died before it was old enough to fly. Spokesmen for the program said that they hadn't expected the eggs to produce a fledgling, given that first attempts like these rarely added up to much.

So when Osborne and raptor expert Chad Olsen of the U.S. Park Service found another active nest cave the next year, they tried to keep their hopes in check. "The egg was in a part of the canyon called the Inferno," Osborne said. "It's a narrow drainage near Hopi Point with incredible red rock walls, and the nest was at the point of the drainage."

Osborne and Olsen hiked down into the Inferno in August of that year. When they got close enough, they set up a sighting scope and looked into the mouth of the cave.

"It was too dark in there to see very much, but Chad thought something was moving. Then he said something like 'Oh my God,' and we saw a very big baby condor come out of the darkness." Osborne said they sat in the Inferno and watched the chick for two more days, sometimes feeling totally cut off from the rest of world and sometimes hearing the voices of the unseen crowds of tourists gathered on the distant observation points.

Out in California, three condor chicks were seen that year. Two died quickly. The third chick appeared to thrive for months, raising hopes that it would fledge. Then, unexpectedly, it started shedding its tail and secondary feathers. The bird was taken out of the cave and flown to the Los Angeles Zoo, but by the time it arrived it was too late. Veterinarians euthanized the chick on September 14. Necropsy results revealed an irreversible lung disease and a hole in the gastrointestinal tract. Wedged into the condor's crop were the pop-tops of three aluminum cans, shards of glass and plastic, and an eighteen-inch-long rag soaked in oil. Critics of the program said the necropsy helped prove that California was no longer safe for condors. Noel Snyder renewed his call for a broad review of the program by a panel of ornithologists with no connection to the program.

"This is why we took the wild condors to the zoos in the first place," Snyder said. "Putting them back into the same environment doesn't make any sense. If we don't reduce the lead threat in particular we'll always have a feeding-station population of condors, and nobody I know wants that."

Osborne and Olsen hiked back into the Inferno in the fall to check up on the lone remaining wild chick, which was now equipped with giant wings it didn't know how to use. "For a while it would have these activity bouts where it would flap its wings madly and run around inside the cave. Then it gave us heart attacks by coming out onto the ledge so it was facing the cliff face and beating its wings against the rocks." The ledge was very narrow and the cave was roughly six hundred feet above the ground.

After that, for a couple of days, the condor barely moved. Osborne and Olsen worried more. On November 5, it had another burst of

energy, jumping up and down while seeming to look in fifty directions at once. Then, after seeming to calm down, it jumped off
the ledge and fell, spinning and tumbling in a way that reminded
Osborne of a maple leaf.

"I don't know if it had something in mind and overestimated its
ability, but it was falling and our hearts were in our throats. The
wings were partially extended and the bird was trying to right itself,
but the most it could manage was a kind of controlled plummet."
After falling about two hundred feet, the condor briefly disappeared
behind a wall of rock that was jutting out of the side of the cliff;
then it tumbled back into view, still out of control, with about four
hundred feet to go before it would crash into the bottom of the
canyon.

Somewhere in that last four hundred feet the condor learned to
fly. Not well, but well enough. "It was a surprisingly gentle landing," Osborne said. "He stood there looking kind of shell-shocked
for a minute or two. Then he started walking toward the nest cave."

NOTES

INTRODUCTION

1. *The town once known as Piru City was built in the 1860s by a man named David Cook*: stories about "Piru City," California, can be found on the Web site of the Santa Clarita Valley Historical Society: www.scvhs.org.

CHAPTER ONE: THE WORST OF TIMES

1. *A condor is [only] five percent feathers, flesh, blood and bone*: open letter to Russell Peterson, National Audubon Society, from David Brower, Friends of the Earth, September 4, 1980.

2. *He went after the basic mathematics underlying the Bureau's proposals and uncovered embarrassing errors*: taken from John McPhee, *Encounters with the Archdruid*, Farrar, Straus and Giroux, 1971.

3. *"Then came my lucky day"*: David Brower's account of finally seeing wild condors is taken from his essay, "Any Places That Are Wild," from "The Uneasy Chair," *Earth Island Journal* (Spring 1987).

CHAPTER TWO: WING IN A GRAVE

1. *The Pleistocene epoch was the condor's prime*: descriptions of Pleistocene scavenging birds devouring the carcass of a mammoth are based in part on Fritz Hertle, "Diversity in Body Size and Feeding Morphology within Past and Present Vulture Assemblages," *Ecology* 75, no. 4 (1944); and in part on Roger Caras, *Source of the Thunder: The Biography of a California Condor*, Little, Brown, 1970.

2. *Is not the California condor a senile species*: taken from Lloyd Miller's essay, "Succession in the Cathartine Dynasty," Univ. California Los Angeles, 1942.

CHAPTER THREE: MORE LIKE RELATIVES

1. *Condors also play a crucial role in the stories:* descriptions of condors with supernatural powers are taken from the following: E. W. Gifford and G. H. Block, "Above Old Man Destroys His First World, as told by the Wiyot Indians of Humboldt County," and "The Making and Destroying of the World as told by the West Mono Indians of Madera County," in *Californian Indian Nights: Stories of the Creation of the World, of Man . . .* , A. L. Clark Co., 1930, repr. Univ. Nebraska Press, 1990; Hubert H. Bancroft, "Myths and Languages," *The Native Races*, vol. 3, A. L. Bancroft & Co., 1882; Dwight D. Simmons, "Interactions Between California Condors and Humans in Prehistoric Far Western America," *Vulture Biology and Management*, Univ. California Berkeley, 1984; passages from Carroll De Wilton Scott, "Looking for California Condors," 1935, 1945, ms. in Henry E. Huntington Library, San Marino, California; A. L. Kroeber, *Indian Myths of South Central California*, Univ. California Berkeley, 1907; E. W. Gifford, *Miwok Cults*, Univ. California Berkeley, 1926; and E. W. Gifford, *Central Miwok Ceremonies*, Univ. California Berkeley, 1955.

2. *But the authors of a recent book on condors take a much more pessimistic view*: rough calculations of the long-term effects of thousands of years of condor sacrifice practiced by Indian tribes are based on Noel Snyder and Helen Snyder, *The California Condor: A Saga of Natural History and Conservation*, Academic Press, 2000.

3. *Not long afterward, an angry group of twenty activists and Indians gathered:* descriptions of the fight between Chumash Indians not recognized by the federal government and the Fish and Wildlife Service are based in part on "Ceremonial Dance for Wild Condors atop Mt. Pinos, June 8, 1987," press release, Earth First!, Earth Island Institute, Committee for Wild Condors, spring 1987; press release, Sept. 3, 1986, Quabajai Chumash Indian Association; Resolution Number 2–86, Native American Heritage Commission, "Memorandum: Chumash Indian/California Condor Program Controversy," California Department of Fish and Game, 1986; letter from Sidney Flores, attorney representing Coastal Chumash Clan, August 1, 1986.

CHAPTER FOUR: SWAY OF KINGDOMS

1. *Off the coast of Terrestrial Paradise, 1602:* imagined description of a scurvy-ridden sailor thinking he sees griffins eating a dead whale on a beach in California is based in part on Harry Harris, "The Annals of *Gymnogyps* to 1900," *The Condor* (Jan. 1941).

2. *Alta California was God's gift:* descriptions of cattle in the Mission and Rancho eras are based in part on L. T. Burcham, *California Range Land: An*

Historico-Ecological Study of the Range Resources of California, Div. of Forestry, State of California, 1957; and Tracy I. Storer and Lloyd P. Tevis, *The California Grizzly,* Univ. California Berkeley, 1995.

3. *Then there were the grizzly bears:* descriptions of grizzly bears and their interactions with condors are based in part on passages from Storer and Tevis, *California Grizzly;* Charles F. Outland, *Mines, Murders and Grizzlies: Tales of California's Ventura Back Country,* 1969, repr. Ventura Co. Museum of History and Art, 1998; and Mariano Guadalupe Vallejo, "Grizzlies at the Calaveras," in *Ranch and Mission Days in Alta California,* repr. in Don DeNevi, ed., *Sketches of Early California,* Chronicle Books, 1971.

4. *There's another key link between the condors and the bears:* accounts of the importance of chaparral to the survival of both grizzlies and condors are based on Charles M. Goethe, *The Elfin Forest: A Glimpse of California's Chaparral,* self-published, 1953; Fred G. Plummer, *Chaparral: Studies in the Dwarf Forests, or Elfin-Wood, of Southern California,* USDA Forest Service, 1911; and Storer and Tevis, *California Grizzly.*

CHAPTER FIVE: COLLATERAL DAMAGE

1. *The California gold rush hit the condor's world like a meteor from the East:* early descriptions of the damage wrought by logging and mining come from miner and artist J. D. Borthwick, as cited by David Beesley in *Crow's Range: An Environmental History of the Sierra Nevada,* Univ. Nevada Press, 2004.

2. *Condors weren't common in the flatlands:* descriptions of the geographic damage done to the Sierra by the gold rush are based largely on "The Sierra Gold Made," Beesley, *Crow's Range.*

3. *"Yonder it descends in a rush of water-like ripples and sweep":* John Muir's musings on the power of the wind after climbing up a tree in a heavy storm are taken from "A Wind Storm in the Forests," *The Mountains of California,* Century Co., 1894.

4. *Condors may also have nested in the trees in these grand forests:* Muir's description of the fires burning through the tops of giant forests in the night are taken from Edwin Way Teale, ed., *The Wilderness World of John Muir,* Houghton Mifflin, 1954.

5. *The market for the wood of these giant trees:* based on J. S. Holliday, *Rush for Riches: Gold Fever and the Making of California,* Univ. California Berkeley, 1999.

6. *One forty-niner described his wagon train as a rolling armory:* Holliday, *Rush for Riches.*

7. *Condors near the mining fields were shot dead for the hell of it:* the account of miner and pony express rider Alonzo Winship sneaking up on a sleeping condor

and whacking it about the head and neck with a shovel comes from M. L. Herman, "The Capture of a California Condor in El Dorado, Colorado," *The Condor* 2, no. 1 (Jan.–Feb. 1900).

8. *Life got even harder for the condors in the 1850s:* broad accounts of the damage done to California's wildlife by "market-hunters" are taken from Holliday, *Rush for Riches*, and Raymond F. Dasman, "Environmental Changes before and after the Gold Rush," in James J. Rawls and Richard J. Orsi, eds., *A Golden State: Mining and Economic Development in Gold Rush California*, Univ. California Berkeley, 1999.

9. *"There were five horses packed with buffalo robes":* accounts of the market hunt during which the famous outdoorsman and bear hunter James "Grizzly" Adams saw his first and last California condor are taken from Theodore Hittell, *The Adventures of James Capen Adams, Mountaineer and Grizzly Bear Hunter of California*, Towne & Bacon, 1860.

10. *California condors got meaner and bigger in the late 1800s:* Alexander Taylor's famously inaccurate description of "The Great Condor of Northwest America" as a carnivorous outlaw bird a cattle or sheep rancher had every right to shoot or poison was initially published in *The California Farmer* (Nov. 1854), and then republished in *Hutchings' California Magazine* in March 1959.

11. *Ornithologist Adolphus Heerman said that scenario repeated itself:* from Pacific Railroad surveys as cited in Harris, "Annals of *Gymnogyps* to 1900."

12. *In most of the state, the grizzlies were in full retreat:* accounts of grizzly hunting in the mountains of south-central California are from Outland, *Mines, Murders and Grizzlies*; Storer and Tevis, *California Grizzly*; Susan Snyder, *Bear in Mind: The California Grizzly*, Heyday Books, 2003; and Allen Kelly, *Bears I Have Met—And Others*, Drexel Biddle Publishers, 1903.

13. *Even the law-abiding citizens seemed to be a little off:* accounts of bandits terrorizing stages being hauled over the top of the pass that separates Southern Caifornia from the Great Central Valley come from John Robinson, "Tiburcio Vasquez in Southern California: The Bandit's Last Hurrah," *California Territorial Quarterly*, no. 1400 (Fall 1996).

14. *Lechler also liked to stress that the first recorded gold strike:* one of the few descriptions of the hollow see-through quill of a foot-long condor feather being used to carry and measure out gold dust comes from John Bidwell, *Life in California before the Gold Discovery*, Lewis Osborne, 1966 (written by Bidwell in the 1890s about the 1840s).

15. *Apparently Cooper hadn't seen the condor for eight years:* the story of ornithologist J. G. Cooper's encounter with a docile condor on a beach in what would later be known as Orange County comes from J. G. Cooper, "A Doomed Bird," *Zoe* 1 (1890).

CHAPTER SIX: SKIN RECORD

1. *When his short, spectacular career began in the early 1800s*: descriptions of John K. Townsend's life rely on the introduction to Townsend's *The Narrative of a Journey Across the Rocky Mountains to the Columbia River*, 1939, repr. Oregon State Univ. Press, 1999.

2. *Thrilled to finally see the bird he'd dreamed of in the flesh*: the story of his fight with a wounded condor in Oregon was told by Townsend himself in "California Vulture," *The Literary Record and Journal of the Linnean Society of Pennsylvania College* 4, no. 12 (Oct. 1848).

3. *No one would have questioned Townsend's need to kill*: books that did a lot to shape my attitudes toward this era include Barbara Mearns and Richard Mearns, *The Bird Collectors*, Academic Press, 1998.

4. *All but one of the condors that were sent overseas left the country as carcasses*: the imagined account of the voyage taken by the only living condor ever to be shipped overseas is based primarily on Harris's "Annals of *Gymnogyps* to 1900."

5. *The only group the scientists did not blame was their own*: accounts of the life and work of Joseph P. Grinnell come from the following sources: Alden Miller, "Joseph Grinnell," *Systematic Zoology* 13, no. 4 (Dec. 30, 1964); Joseph Grinnell, "Old Fort Tejon," *The Condor* 7 (Jan.–Feb. 1905); Joseph Grinnell, *Joseph Grinnell's Philosophy of Nature: Selected Writings of a Western Naturalist*, Univ. California Press, 1943.

6. *California's condor owners said their birds could do all that and more*: stories about pet condors, especially the bird known as the General, come from the following sources: Frank H. Holmes, "A Pet Condor," *The Nidologist* (date unknown); William L. Finley, "Home Life of the California Condor," *The Century Magazine* 75, no. 3 (Jan. 1908); William L. Finley, "Life History of the California Condor," *The Condor* 10, no. 1 (Jan.–Feb. 1908); and William L. Finley, "The Passing of the California Condor," *The Condor* (1926).

CHAPTER SEVEN: EGGMEN

1. *I have a picture of an egg collector on my desk*: most of what I know about eggs of condors I learned from Lloyd Kiff, a former oologist who now works for the Peregrine Fund. Kiff very kindly lent me the unpublished studies he has done of almost every collected condor egg now in existence.

2. *A flat wooden tray full of birds' eggs rests in Kelly Truesdale's lap*: the rest of what I know about the egg collector Kelly Truesdale comes from Ian McMillan, *Man and the California Condor*, E. P. Dutton, 1968, or from William Dawson's account of a condor egg–collecting trip he took with Truesdale once; that account is part of Dawson's massive study, *The Birds of California*, South Moulton Company of Los Angeles, 1923.

3. *Kelly Truesdale starred in some of the best pulp nonfiction:* accounts of some of the crazy things egg collectors used to do are taken mostly from Joseph Kastner, *A World of Watchers: An Informal History of the American Passion for Birds,* Sierra Club Books, 1986; articles that also shaped my thoughts included Harry H. Dunn, "How I Found the Nest of a Condor," *The American Boy* (Feb. 1907) and H. G. Rising, "The Capture of a California Condor," *The Bulletin of the Cooper Ornithological Club* 1, no. 2 (Spring 1889).

4. *"And as he drove back to headquarters":* the quote is from Earle Crow, *Men of El Tejon,* Ward Ritchie Press, 1957.

5. *Everyone who's ever lived in these small towns:* the story of the St. Francis dam disaster is best told in Charles F. Outland's *Man-Made Disaster: The Story of St. Francis Dam,* A. H. Clark Co., 1963.

CHAPTER EIGHT: PATRON SAINT

1. *The best thing that can happen:* introductory quote and others that clearly follow are from H. H. Sheldon, "What Price Condor?" *Field & Stream* (Sept. 1939).

2. *The jolting launch makes the pilot and crew of the plane:* quotes I imagine Koford thinking while he's flying a scout plane in World War II are based on Carl Koford, *The California Condor,* Dover, 1953.

3. *Carl Koford is the patron saint of condor field research:* notes taken by Koford in the field are transcribed from a copy of his field notes kept at the Museum of Vertebrate Zoology at the University of California at Berkeley.

4. *Koford hitched his first ride into condor country in the spring of 1939:* the letters of introduction from Joseph Grinnell that Koford carried in the field were supplied by the Museum of Vertebrate Zoology at the University of California at Berkeley.

5. *"Carl was frugal":* Ian McMillan's description of Koford's beat-up car comes from McMillan, *Man and the California Condor.*

6. *These two men were Koford's best friends in the field at the time:* allegations that Koford often handled condors in their nests before the start of World War II are based on films the late Ed Harrison showed me in his office, and photographs Koford took of Harrison in the back of a cave at the Sespe Condor Sanctuary, holding the condor chick known as Oscar.

7. *This was not a notion that went over very well in the scientific community:* my views of Koford's complicated legacy were shaped by an unpublished manuscript provided by Roland Clement, former science director at the National Audubon Society; by Snyder and Snyder, *The California Condor;* and by the text of the only lengthy interview Carl Koford ever sat for, in David Phillips

and Hugh Nash, eds., *The Condor Question: Captive or Forever Free?* Friends of the Earth, 1981.

CHAPTER NINE: HANDS-ON

1. *Ian McMillan was removed from this committee:* McMillan's statement about being kicked off the scientific review panel set up to monitor Sibley's work comes from an interview with McMillan published in Phillips and Nash, *The Condor Question.*

2. *Sibley and McMillan never got the chance to talk about becoming allies:* accounts of Sibley's fieldwork and his struggles to get along with hands-off condor activists, the U.S. Forest Service, and local business groups pushing for permission to dam the river that runs through the condor's breeding grounds are based partly on e-mail exchanges, but mostly on the regular reports Sibley sent to the Endangered Species Wildlife Station at the USGS Patuxent Wildlife Research Center in Maryland.

3. *Topa Topa hasn't seen the wild since:* some of the descriptions of Topa Topa misbehaving at the Los Angeles Zoo come from the unpublished "Notes on the Behavior of Topa Topa," written for the zoo by former condor keeper Frank Todd in 1971.

4. *The Sespe Creek project skidded off the tracks in the summer of 1967:* McMillan's claim that the public vote against building the Sespe dam project was a sign of the environmental times comes from his book *Man and the California Condor.*

5. *McMillan and his allies failed to thank Fred Sibley for helping to nail this coffin shut:* Sibley's claim that the dam will drive the condor extinct is made in a report entitled "Effects of the Sespe Creek Project on the California Condor," August 1969.

6. *Sibley quit his job as a condor biologist:* accounts of how Fred Sibley lost his job come from numerous sources.

CHAPTER TEN: CONTINGENCIES

1. *A story on the Internet started me down this path of inquiry:* Dave Boehi, "Mr. Wilbur Loves the Condor" can be found online at www.wwcmagazine.org, which is affiliated with Campus Crusade for Christ.

2. *Wilbur defined these drastic steps in a Contingency Plan:* references to the fight over whether to permit phosphate mining in the condor refuge are based on various news reports and an internal memo on an August 3, 1970, meeting involving various government agencies and representatives of the mining industry and the National Audubon Society.

3. *But the activists were outraged by the "last-ditch" actions:* reports on the internal debate over whether to include a last-ditch captive breeding plan in the updated version of the California Condor Recovery Plan required by the U.S. Endangered Species Act are described at length by Sanford R. Wilbur in his self-published *Condor Tales: What I Learned in Twelve Years with the Big Birds*, 2004.

4. *"The existence of the California condor depends on conscientious human intervention":* quotes from the panel of "independent" scientists asked to review the condor's status and recommend conservation priorities are taken from "Audubon Conservation Report No. 6: Report of the Advisory Panel on the California Condor," June 1978.

CHAPTER ELEVEN: ENDGAME

1. *A scientific SWAT team rolled into the Sespe in the summer of 1980:* accounts of the events that led to the capture of the last wild condor in 1987 are as numerous as they are varied. Noel Snyder's point of view is laid out quite extensively in *The California Condor*; David Darlington, a hands-off activist, profiled the legendary activist/ranchers Eben and Ian McMillan in his book *In Condor Country*, Henry Holt, 1987. Other accounts can be found in Phillips and Nash, *The Condor Question* and Wilbur's *Condor Tales*.

2. *Snyder was right about the activists:* Eben McMillan's letter to Governor Edmund G. Brown was found in a filing cabinet at a U.S. Fish and Wildlife Service regional office in Ventura, California.

3. *Brower and Phillips asked to see the film of the disaster:* David Brower's side of the arguments that followed the accidental killing of a condor chick in its nest by a USFWS biologist are recounted in Phillips and Nash, *The Condor Question*, which includes written statements between Dave Phillips of Friends of the Earth and Bill Lehman, the government biologist in whose arms the chick died.

4. *Brower and his colleagues jumped all over that claim:* open exchanges between Brower and Lehman and a telegram Brower sent to Interior Secretary Cecil Andrus are also published in Phillips and Nash, *The Condor Question*.

5. *The California Condor Recovery team did a huge amount of living on condor time:* accounts of what the field teams did until the state permits that allowed them to trap condors, fit them with radio transmitters, and ultimately take them to the San Diego Wild Animal Park and the Los Angeles Zoo are based in part on field notes taken by the members of the condor field teams.

6. *Condor country had become a risky place for condors:* Audubon's reluctance to allow all the birds to be taken into captivity is justified in an internal memo sent to some of the leaders of the condor program.

7. *More ugliness followed:* exchanges between Marsha Hobbs of the Los Angeles Zoo and Peter Berle of the National Audubon Society are based on dueling press releases and news reports.

CHAPTER TWELVE: ZOO

1. *The condor in the kennel behind the front seat:* almost all this chapter is based on interviews with scientists and others at the Los Angeles Zoo, the San Diego Wild Animal Park, the Center for Reproduction of Endangered Species at the San Diego Zoo, and the people who work in the Captive Breeding Center at the Peregrine Fund. But my opinions were no doubt shaped indirectly by the following written materials: Janny Scott, "An Attractive Bird? Condor Passes Health; Mate to be Picked," *Los Angeles Times*, April 21, 1987; Nancy Ray, "Ambitious Plan Is Last Hope to Rescue Bird from Extinction," *Los Angeles Times*, March 25, 1982; William D. Toone, "Rodan Revisited," *Zoo News*, San Diego Zoo, December 1981; "Cathartid Feeding Protocol," 1985 memo, Los Angeles Zoo; Michael Soule and J. Verner, "Population goals for the California Condor," unpublished report, March 16, 1987; and Sanford Wilbur and Jerome Jackson, eds., *Vulture Biology and Management*, Univ. California Berkeley, 1983.

2. *The point of the article was in the headline itself:* Bil Gilbert, "Why Don't We Pull the Plug on the Condor and Ferret?" *Discover* (July 1986).

3. *Kiff had signed off on the recovery plan:* the legal claim that the California condor has been "technically extinct" since the last free-flying bird was captured in 1987 has never been pursued by the Tejon Ranch, but I'm told that it is still on file at the USFWS Sacramento office.

4. *In 1989, these gathering land-use fights were postponed by a controversial experiment:* accounts of the debate over whether Andean condors should be released "to hold the habitat" until the California birds were ready to return are found in Wilbur, *Condor Tales*; Snyder and Snyder, *The California Condor*; and various news reports, including David Smollar, "Can Andean Condor Help Save California Kin?" *Los Angeles Times*, Nov. 7, 1986.

5. *Early April, 1988:* some of what I know about the first hatching of a California condor egg laid in captivity comes from "Condor Chick Stuck in Egg, Scientists Stand by to Help," wire reports, *Los Angeles Times*, April 30, 1988; and "Healthy Condor Hatches in San Diego," *San Francisco Chronicle*, April 30, 1988.

CHAPTER THIRTEEN: GRAND CANYON

1. *When Robert Mesta saw himself hanging in effigy:* Robert Mesta's account of the disastrous meeting in Kanab is confirmed by news accounts and angry letters to the editors of various local papers.

2. *This was the mood in Fredonia when the condor plan was officially unveiled in 1995:* the writing of this chapter was also shaped by the Peregrine Fund's "A Review of the First Five Years of the California Condor Reintroduction Program"; Christopher Woods, Shawn Farry, and William Henrich, "Survival of Juvenile and Subadult Condors Released in Arizona," unpublished, 2001; and by the arguments of Vicky J. Meretsky, Steven Beissinger, Noel Snyder, David Clendenen, and James Wiley, "The California Condor: A Flagship Adrift," *Conservation Biology* (August 2000).

3. *New groups of condors were released on a regular basis:* Shawn Farry's accounts of the incredible amounts of work involved in managing condors match what he wrote in his field notes.

CHAPTER FOURTEEN: NOT THE SAME BIRD

1. *Farry paused to explain why the friendly condors were the most at risk:* accounts of condors landing on everything from power lines to airport runways were gleaned from various news reports and interviews with representatives of the California Condor Recovery program.

2. *Burnett said he'd been worrying about these birds since the day they arrived:* efforts to "train" the birds to act like "wild" condors were gathered the same way as above.

3. *Snyder's findings came as no surprise:* reactions inside the condor program to the publication of Meretsky et al., "The California Condor: A Flagship Adrift," and then to Snyder and Snyder, *The California Condor,* are described in unpublished letters to the editor of the journal *Conservation Biology.*

CHAPTER FIFTEEN: THE REAL KILLERS

1. *The energy company that wanted to build this was called Enron:* my account of the fight over whether to allow the Enron Corporation to build so-called condor Cuisinarts in the middle of the condors' feeding range are based on various news accounts and press releases and on interviews with representatives of the National Audubon Society and the Tejon Ranch. I was unable to reach anyone who would agree to speak for Enron.

2. *Snyder says the problem is lead shot in the carcasses of animals:* the crises that ensued when at least six California condors contracted potentially lethal cases of lead poisoning after eating from what may have been a single but unknown food source was described at length by Farry in unpublished field notes and sharply criticized by Snyder and his allies in news reports and interviews.

3. *So what were the clues?:* studies that should have raised alarms over the possible links between the use of lead shot by hunters and the unexplained deaths of

condors have been in play since at least late 1986, when O. H. Pattee, P. H. Bloom, J. Michael Scott, and Milton R. Smith published "Lead Hazards with the Range of the California Condor," *The Condor* 92. More recent studies include V. J. Meretsky et al., "Demography of the California Condor: Implications for Reestablishment," *Conservation Biology* (2000); and Michael Fry, "Assessment of Lead Contamination Sources Exposing California Condors," submitted to the California Department of Fish and Game, April 7, 2003.

CHAPTER SIXTEEN: ELVIS REENTERS THE BUILDING

1. *The plan to put Igor back where he had come from was approved in December 2002:* interviews conducted at the California Condor Recovery team meeting at the Los Angeles Zoo in December 2001 were augmented by extensive written materials distributed at that meeting.